Advanced Level Technical Drawing

E. JACKSON, M.COLL.H.

 Longmans

Longmans, Green and Co Ltd

London and Harlow

*Associated companies, branches and representatives
throughout the world*

© Longmans, Green and Co Ltd 1968
First published 1968
Second impression 1969

SBN 582 32326 6

Set in Monophoto Times New Roman and made and printed by offset in
Great Britain by William Clowes and Sons, Limited, London and Beccles

Contents

Foreword

This book is a comprehensive textbook covering 'A' Level Examinations in Geometrical and Engineering Drawing and should be suitable for both individual study and classroom work. Metric units have been introduced in anticipation of the coming changes in examination requirements.

The first part of the book deals with plane geometry, conics, involutes and gearing, cams and mechanisms, centroids, vector polygons and frameworks.

The projection of a point, line, laminae and solid to the inclined and oblique plane is progressively treated, and interpenetrations of solids are shown, from simple examples to angled offset cone, prism, cylinder and sphere. Developments by simple projection and by triangulation are also covered.

The second part deals with engineering components and conventional drawing practice. Following present examination trends, the machine drawing examples are shown as parts to be assembled, from which sectioned and full views are required in either first or third angle projection. Isometric cutaway drawings of the assemblies are shown to assist in the explanations, examples in pictorial projection using perspective, oblique and axonometric views are given.

Worked examples and exercises are given in all sections of the book, from fundamental principles to problems of examination standard, using upwards of a thousand diagrams.

EDWARD JACKSON

Acknowledgements

We are grateful to the following for permission to reproduce copyright material:

The Associated Examining Board for the General Certificate of Education, the Joint Matriculation Board (Universities of Manchester, Liverpool, Leeds, Sheffield and Birmingham) and the University of London for questions from past examination papers, and the British Standards Institution for extracts from B.S. Specifications.

Part I
GEOMETRICAL DRAWING

Plane Geometry

1. Proportionate Division by intersecting parallels. Draw the two lines at a convenient angle, join ends, and by parallels obtain the proportionate division of the new line at A′, B′, C′, D′.

2. Rectangles of Equal Area Extend base to new length at A. Erect perpendicular to height of original rectangle. Draw the diagonal, and through B draw a parallel to the base line. The construction may be reversed.

3. Triangles of Equal Area Enclose triangle in rectangle. Extend base to new length, and obtain enclosing rectangle for new triangle. Draw new triangle in new rectangle, several solutions, using same base and height.

4. Square Roots Draw a right-angled triangle, two sides indicated one unit each. Hypotenuse gives root 2. Continue the construction with further right-angled triangles to obtain roots 3, 4, 5 etc. by scale.

5. Logarithmic Progression Given the ratio and angle between two lines, draw the lines enclosing the angle and join the ends of the lines. Draw the arc to give position 1, and draw further arcs and parallels to obtain further points and lengths. These are used later in the construction of the Logarithmic Spiral, see 110.

6. Circles Touching Lines and Circle Common Tangency. Draw the two given lines. Obtain centre of first circle from intersection of lines parallel to given lines, distance R_1. Centre of second circle is at intersection of arc radius $R_1 + R_2$ from first centre and line parallel to base line, distance R_2. The common tangent to the circles is also shown. This construction is also useful in diagrams showing sections of touching spheres.

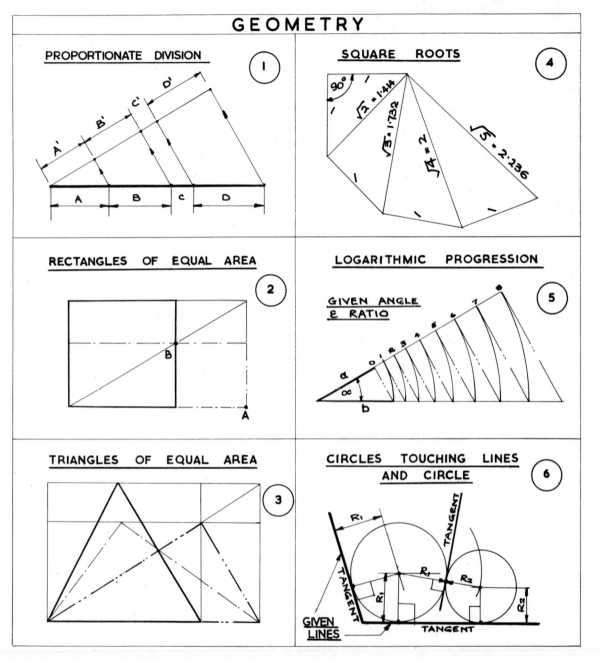

GEOMETRY

PROPORTIONATE DIVISION ①

SQUARE ROOTS ④

RECTANGLES OF EQUAL AREA ②

LOGARITHMIC PROGRESSION ⑤

GIVEN ANGLE & RATIO

TRIANGLES OF EQUAL AREA ③

CIRCLES TOUCHING LINES AND CIRCLE ⑥

7. Touching Circles Deduct radii for internal circle and add for externally touching circle. Point of contact on line joining centres, tangent at right angles to this line.

8. Three Touching Circles $R_1 + R_2$, $R_1 + R_3$, and $R_2 + R_3$ give the three sides of a triangle. Apice are centres for touching circles. Conversely, given the triangle, three touching circles may be found, the radii being the length of the perpendicular from each side intersecting the bisectors of the angles of the triangle.

9. Straight Line Equal to Arc Draw the arc and sector. Extend the chord by half. Draw the tangent at A, and cut off by an arc from B. Length required is AC.

10. Area by Ordinates Simpson's Rule, as shown. The simpler method is to add the length of mid-ordinates of each strip and multiply by the width of a strip by number of strips.

11. Regular Polygon Given the length of a side, draw the semicircle on it, and divide into as many parts as sides. Join always point 2 to centre. Bisect the sides formed, and from the centre obtained draw the enclosing circle, step off the remaining sides.

12. Regular Polygon inside a given circle. Draw diameter and from its ends draw intersecting arcs, radius as diameter. Divide the diameter into as many parts as sides of the polygon, a line drawn through 2 from A cuts the circle giving the length of the first side of the polygon.

GEOMETRY

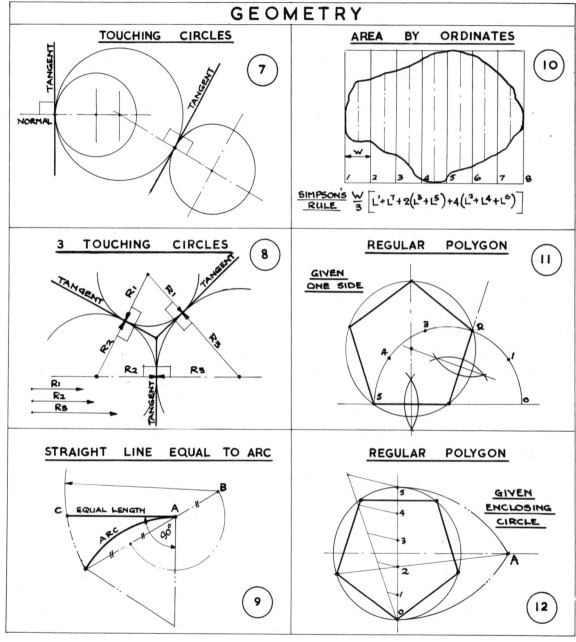

Simpson's Rule: $\dfrac{W}{3}\left[L^1 + L^7 + 2(L^3 + L^5) + 4(L^2 + L^4 + L^6) \right]$

13. Arc Equal to Straight Line (approx.). Draw the circle given the radius of the arc. Draw the line tangential. Divide the line into four equal parts and with centre 1 and radius 1–4, draw arc to cut circle.

14. Circle to Touch Lines and Point All circles tangential to the lines have centres on the bisector of the angle. The line passing through P gives similar points P on other circles. The dotted lines joining centre and P for each circle are parallel.

   A construction circle (two shown) should be drawn first and a line parallel to P_1, O_1, passing through P gives the centre of the required circle, O.

15. Circle to Touch Point and Circle Join P to the centre of the circle and extend to cut the circumference in Q. PQ is the diameter of the circle required. A second solution with A as centre may be drawn.

16. Circle to Touch Point and Circle Draw the construction circle centre P and radius as the given circle. Obtain Q by drawing the perpendicular. Complete the construction as in the diagram.

17. Common Tangents for Touching Circles Given the centres of two circles A and B, radius R_1 and R_2, draw the arcs of touching circles radius R_3 and R_4, as shown. Obtain the centre X of the outer touching circle by subtraction of radii, and then drawing intersecting arcs from A and B. Obtain the centre Y of the inner touching circle by addition of radii, and drawing intersecting arcs from A and B. The points of common tangency are found by joining centres and extending the lines to cut the circumference. The tangents are drawn normal (perpendicular) to the radius lines at the tangent points.

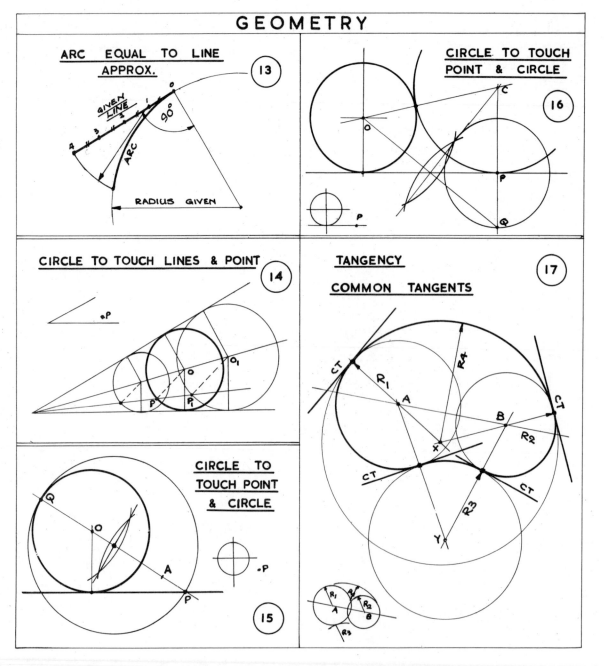

18. Tangency A circle drawn tangential to two lines which meet. The centre will lie on the bisector of the angle, also on intersection of lines which are radius-distant and parallel to the lines. These lines are the locus of the centres of such circles which roll on the lines inside the angle.

19. Tangency The locus of the centre of a circle of given radius rolling inside a given shape.

20. Tangency The diagram shows a shape formed by circles and straight lines. The point of common tangency is given by intersection of the lines joining the centres of the circles concerned.

21. Tangency The locus of the centre of a circle rolling on a given line is shown. Notice how the circle passes the re-entrant and salient corners, and how this affects the locus line.

22. Approximate Compass Curve for an Involute Curve. This construction may be used in the representation of involute gear teeth, when an arc approximate method is allowed. Draw the involute by the usual tangential method where the tangent is made equal in length to its respective arc. Draw a chord in the working length of the involute, and draw a perpendicular to the mid-point of the chord. The centre of the approximate arc is C. Points C for the teeth will lie on a circle radius CO.

23. Enlargement by Radials Enlarge the given shape in ratio 4:5. Draw the given shape. Draw radial lines through the critical points of the shape from pole O.

Enlarge the base of the shape by intersecting parallels. The centres of arcs can be found by radials passing through the first centres intersected by lines from new positions, as shown at C.

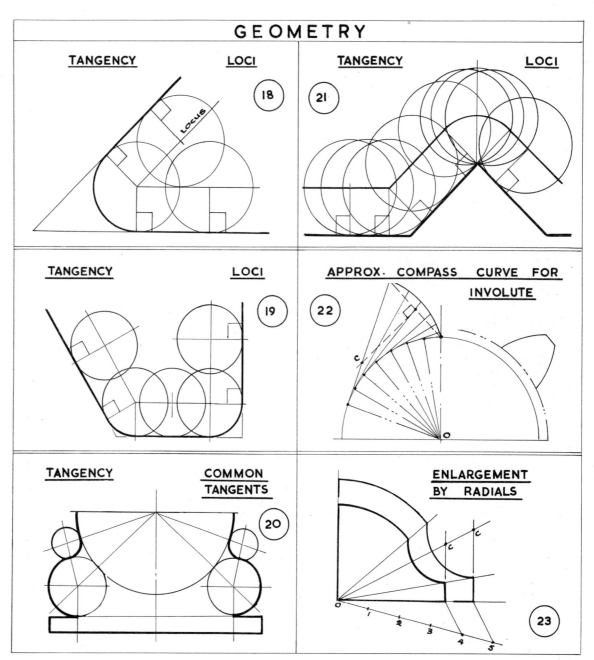

GEOMETRY

5

Exercises

1. Draw a line 5″ in length. Divide in the ratio of 1:2:3:4:5. Divide another line $7\frac{3}{8}$″ long from the first.

2. Given a rectangle 10 cm × 7·3 cm, draw a rectangle equal in area but having a base of 6·5 cm.

3. Given a triangle of sides in the ratio of $2:3\frac{5}{8}:1\frac{1}{2}$, and the length of the first side $1\frac{7}{8}$″, draw a triangle equal in area but with a base of $4\frac{1}{4}$″.

4. Prove graphically that the square root of 5 is 2·236.

5. Draw a logarithmic progression given an angle of 50° and two vectors in the ratio of $1\frac{1}{4}:1\frac{5}{8}$.

6. Two circles, 7 cm and 4 cm diam., touch each other and the smaller touches two lines which are at 85°. Draw the diagram, showing several positions of the larger circle as it moves from contact with one line to the other line. Plot the locus of a point on the large circle during the movement.

7. A circle of 2″ diam. stands on a straight line. A second circle, $2\frac{1}{4}$″ diam., touches the line and the first circle, then rolls round the outside of the first circle until it comes to rest at the straight line. Plot the locus of a point on the moving circle which begins in contact with the straight line.

8. Three circles, 3·3 cm, 5·5 cm, 6·6 cm in diam. lie touching each other. Draw the construction.

9. Draw the arc shown by a radius of $3\frac{1}{4}$″ and subtended by an angle of 75°. Draw a line equal in length to the arc.

10. Draw an irregular shape similar to No. 10, in a rectangle 9 cm by 7 cm. Find the area by Simpson's Rule and also by simple ordinates; compare the results.

11. Construct a regular nonagon inside a circle of 4″ diam.

12. Construct a regular polygon of seven sides, each side 37 mm.

13. A straight line is $3\frac{7}{8}$″ long. Cut off graphically, an arc of a circle $5\frac{7}{8}$″ diam., and equal in length to the straight line.

14. Two lines contain an angle of $82\frac{1}{2}$°. A point P lies $1\frac{1}{4}$″ from the intersection and $\frac{1}{2}$″ from one of the lines. Draw a circle to touch the two lines and pass through P.

15. Repeat the construction shown in No. 15, O Q to be 25 mm.

16. Draw a circle $2\frac{1}{4}$″ diam. tangential to a straight line. A point P lies on the line $2\frac{3}{4}$″ from the point of tangency. Draw a circle to touch the given circle and passing through P.

17. A line A B is 50 mm long. A and B are centres for circles of 65 mm and 40 mm diam. A circle, of 45 mm diam., rolls without slipping round the outside of the first two circles. Plot the locus of a point on the rolling circle.

18. A rolling circle of 2″ diam. rolls without slipping along a straight line 2″ long, then up a slope $2\frac{1}{4}$″ long and gradient of 1:3, then down a slope $1\frac{1}{2}$″ long and at 45° to the original line. Draw dominant positions of the rolling circle from which the locus of the centre point may be seen. Draw the locus of a point on the rolling circle beginning at the point of tangency with the first line.

19. Take the shape drawn in No. 10, and enlarge by radials in the ratio of $4\frac{1}{8}:5\frac{1}{4}$.

20. Two circles, 30 mm and 45 mm diam. are tangential. A third circle, 85 mm in diam., encloses the other two circles and touches both. Draw the figure.

ORTHOGRAPHIC PROJECTION

First Angle

Third Angle

PICTORIAL PROJECTION:

Isometric

Oblique

Axonometric

Perspective

24. First Angle Projection The object is visualised as being placed in a box with a bottom and three sides. The bottom represents the horizontal plane on to which is projected the plan. The three sides represent the three vertical planes on to which the front elevation (centre) and two end elevations (ends) are projected. The isometric explanatory view should help to make this clear.

A typical drawing layout is shown; the plan lies below the front elevation, the end elevations are projected 'across' the front elevation and are views of the 'opposite' end of the object.

To begin the drawing, the views required should be noted, the overall sizes calculated for each view, and the position of the X Y line fixed.

The front elevation and plan are usually drawn in related projected positions, and the end elevations then projected. Special details may require special treatment in projection. A centre line in either plan or elevation, or a shaft or a boss centre may form the dominant point from which a start is made.

'Heights remain the same in all elevations' and are easily projected by the tee square.

Points in the plan can be returned to front elevation by simple projection. Points in the plan can be returned to the end elevations by using either the 45° setsquare or compass.

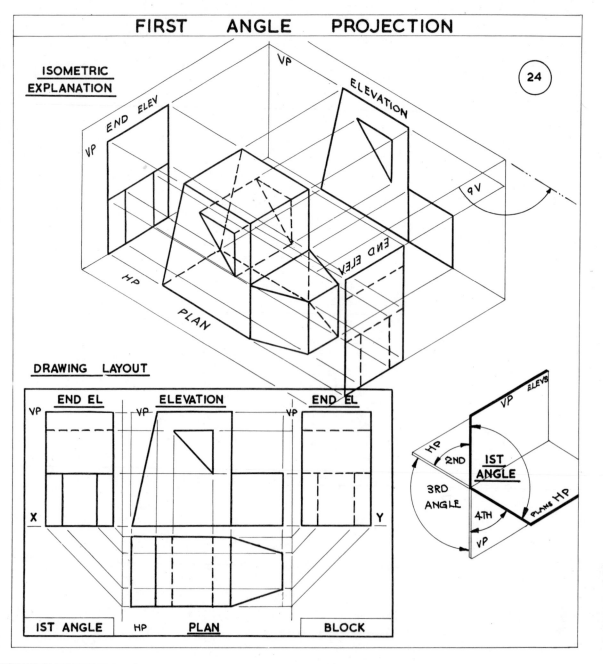

FIRST ANGLE PROJECTION

ISOMETRIC EXPLANATION

DRAWING LAYOUT

25. Third Angle Projection In this projection the object is placed in the third angle shown in the small sketch. The plan and elevations are projected on to the planes which are visualised as being transparent, the object being viewed through them. When the planes forming the transparent box are opened out, the plan appears above the front elevation, whilst the end elevations are diagrams of the ends nearest to the plane and not of the opposite end as in first angle projection. The student is required to have a knowledge of both projections, and to be able to interpret drawings set in either. All the objects and exercises can be projected in first or third angle, and the projection should be stated.

The drawing procedure is similar to that of the first angle. Note the views to be drawn, calculate the overall sizes of the object, and so the space required for each view. From this information, the position of the X Y line may be fixed and the front elevation and plan begun. A centre line forms the usual starting point, or shaft or pulley centre, from which measurements may be taken. The end elevations can be projected from the plan and front elevation by either 45° setsquare or compasses.

More complicated layouts with lists of parts are given in later drawings.

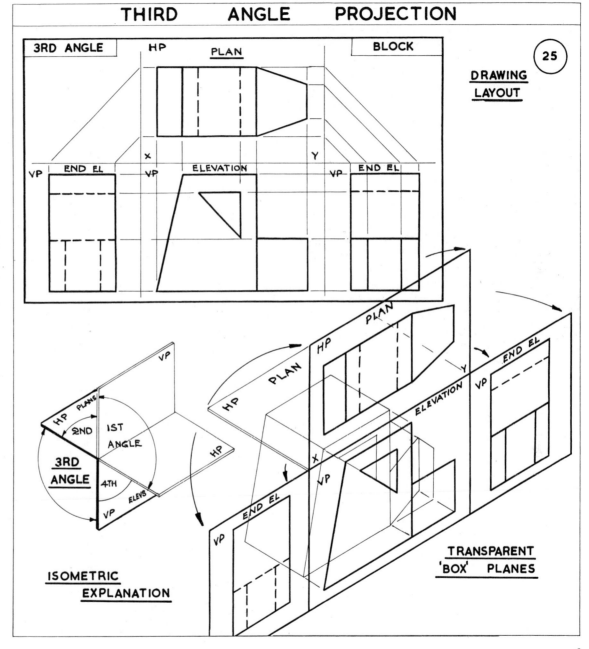

26. Isometric Projection A method of pictorial representation on three planes arranged at 120° to each other, easily constructed by the 60° setsquare. Measurements can only be made on the three axes or their parallels. When strict Isometric Projection is asked for, natural scale measurements are diminished as in the diagram from the 45° line to the 30° line. These measurements are used on the left and right axes A and B to give a pictorial foreshortening. In ordinary isometric drawing—which is commonly used—the ordinary scale sizes are used on the axes. Note that sizes can only be used on the axes or their parallels, since the diagonals of the face are not equal.

The isometric drawing is best begun by drawing the enclosing 'box' sizes obtained from the orthographic projections, plan and elevations. Simple rectilinear shapes can be drawn by projecting points on the faces of the box from measurements on the axes. Two ordinates will intersect to find a point or series of points.

27. Shaped Block Simple shapes and slopes are shown in the exercise shown, and full construction lines are given.

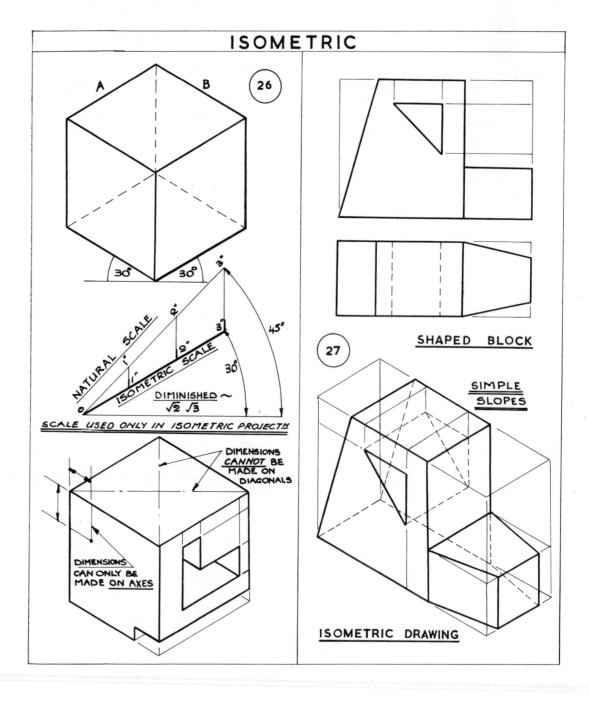

ISOMETRIC

SHAPED BLOCK

SIMPLE SLOPES

ISOMETRIC DRAWING

10

28. Compound Slopes Slopes shown in the figure will require projections from two ordinates before the point in depth can be found to which an internal diagonal line can be joined. Such a point is shown at A in the front face of the object. The square window opening in the face is drawn by using a centre line and ordinates from the face of the 'box'.

29. Curves in Isometric Compass or freehand curves of an object such as the bracket are plotted by ordinates to the curve from the axes. These are drawn in the elevation first and the measurements transferred to the isometric 'box', where the ordinate is marked to the corresponding length. A fair curve is drawn through the points.

The projection of a circle in elevation to isometric can be effected by ordinates, four ordinates and four points on centre lines being the minimum.

The Four Arc Compass Method enables ellipses to be drawn more rapidly but approximately, the centres being easily obtained by cutting the centrelines of the face by lines joining the ends of the centre lines to the opposite corner as shown. Four arcs using two radii allow the projected circle to be executed neatly, and is especially useful in engineering diagrams which employ many cylinders.

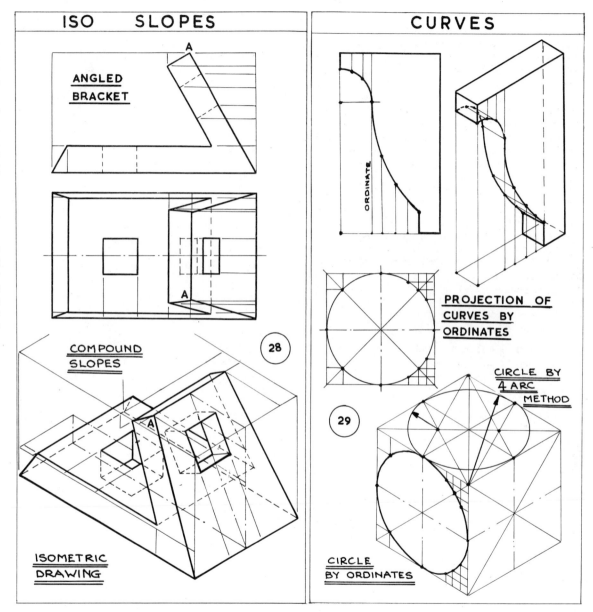

ISO SLOPES

ANGLED BRACKET

A

A

COMPOUND SLOPES

28

A

ISOMETRIC DRAWING

CURVES

ORDINATE

PROJECTION OF CURVES BY ORDINATES

29

CIRCLE BY 4 ARC METHOD

CIRCLE BY ORDINATES

11

30. Compound Forms in Isometric Drawing Difficult forms are best considered as being built up of simple geometric solids, an enclosing 'box' being drawn as part of the basic construction.

In the sectioned bearing block shown, a complete enclosing box should be first drawn, the flange in the foreground being easily drawn by simple ordinates. The cylinder can be drawn using either ordinates to give the ellipse, or by using the quick four arc method previously shown. Having drawn the curves on the front face, the centres may be projected at 30° to the thickness of the block and the remainder of the rear face curves drawn. The diagram should make clear the construction when part of the bearing is removed and a sectioned face exposed.

31. Cone and Pyramid The enclosing box is drawn, the base shape of the circle or polygon constructed. A centre line, vertical in these cases, helps in drawing the top face, a point only for the cone, a small polygon for the pyramid. In the special case of the sphere, a circle drawn to simple measurements would appear too small, and the radius of the circle is usually increased in the reverse order to the isometric scale shown earlier. The larger radius is used from the original base centre.

32. Geometric Solids, and moulded forms. In the eight examples shown, construction lines show how the forms have been built up. An enclosing 'box' may be used in all cases and the simpler forms drawn using ordinates. Alternatively, the drawing may be started from the basic 'ground' shape, ellipse in the case of the cone for example, and a centre line erected therefrom, enabling the rest of the construction to be made.

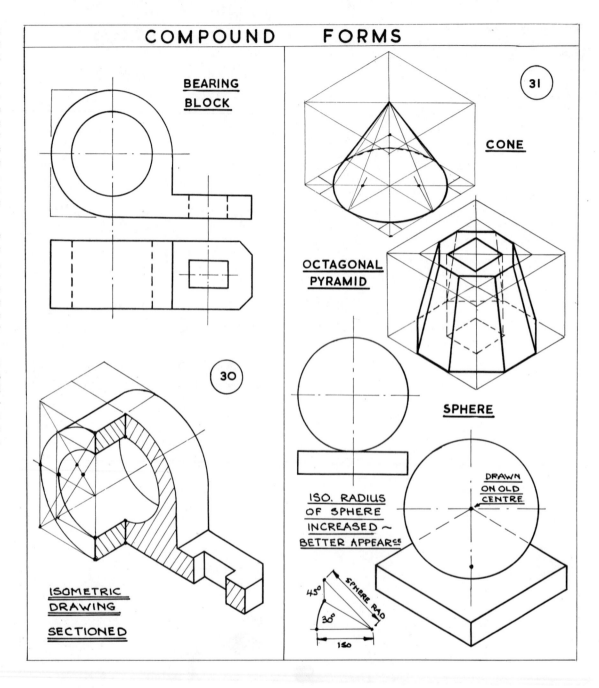

COMPOUND FORMS

BEARING BLOCK

CONE

OCTAGONAL PYRAMID

SPHERE

30

31

ISOMETRIC DRAWING SECTIONED

ISO. RADIUS OF SPHERE INCREASED ~ BETTER APPEAR^{CE}

DRAWN ON OLD CENTRE.

45° 30° SPHERE RAD ISO

COMPOUND FORMS IN ISOMETRIC

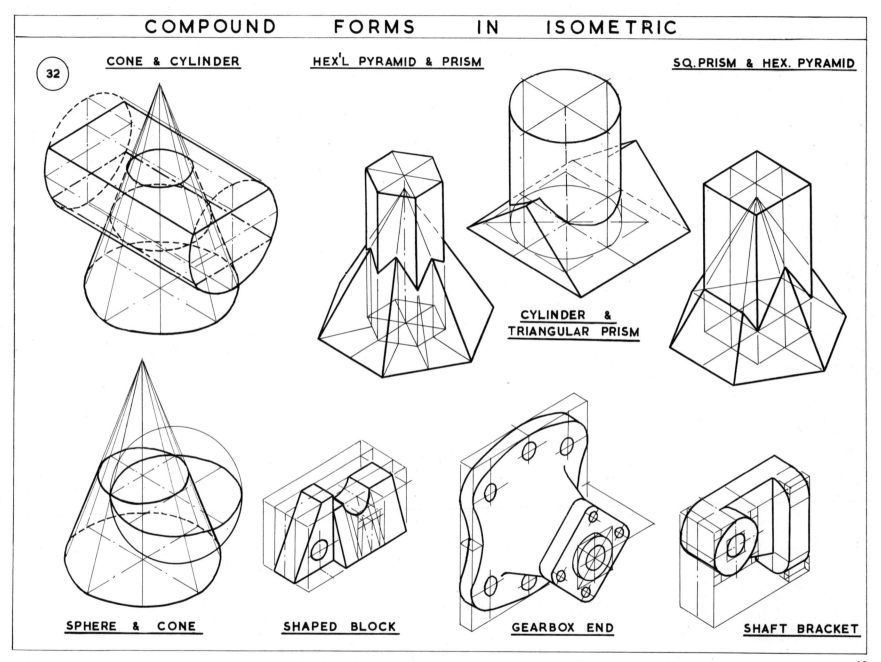

CONE & CYLINDER

HEX'L PYRAMID & PRISM

SQ. PRISM & HEX. PYRAMID

32

CYLINDER & TRIANGULAR PRISM

SPHERE & CONE

SHAPED BLOCK

GEARBOX END

SHAFT BRACKET

13

33. Oblique Projection In this pictorial projection, three axes are used, two at right angles to each other, the third at 45° or 30°. The elevation is usually taken as the front face, the depth of the object being shown on third axis. A cube is shown in Fig. 1, full size measurements on the depth make the object appear distorted (cavalier projection), and half depths (cabinet projection), or even $\frac{3}{4}$ scale is used to give a more proportionate diagram. A series of simple solids are shown in oblique projection, the shapes bounded by straight lines.

Block A shaped block having sloping sides. Draw the enclosing 'box'; measurements must be made only on the axes.

Block. 2nd Position The enclosing 'box' may be drawn with the third axis pointing downward either left or right, and the object presented as shown. This angle of viewing may be used to advantage where details of the bottom face have to be shown.

Pentagonal Prism on base. Simple solids are easily projected by employing enclosing 'boxes'; a centre line may help in drawing objects whose front face does not lie wholly in the same plane.

Octagonal Pyramid Draw the base octagon in oblique projection, erect a centre line to the correct height and construct the upper octagon. Join the corners of the octagons.

Square Prism and Pyramid Begin the interpenetration by constructing the pyramid. From the elevations, obtain height positions of the prism and intersection points. Project the prism from centre lines as shown. Project intersection points and join to complete the drawing.

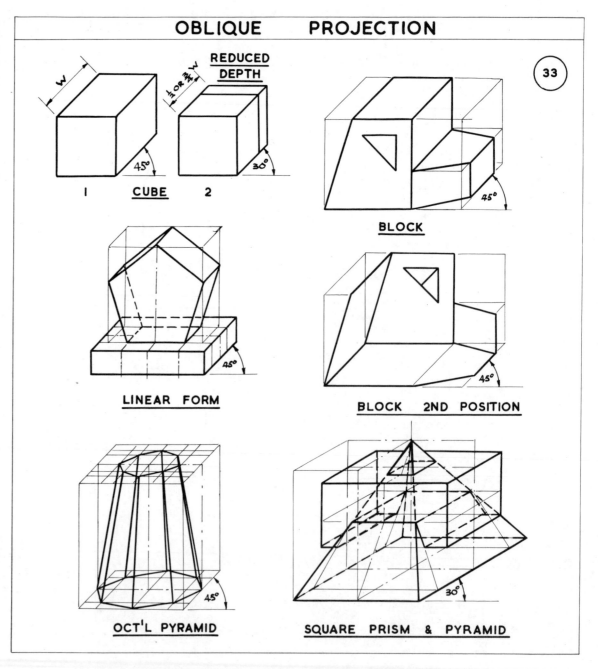

OBLIQUE   PROJECTION

33

REDUCED DEPTH

45°   30°

1   CUBE   2

BLOCK

LINEAR FORM

BLOCK   2ND POSITION

OCT'L PYRAMID

SQUARE PRISM & PYRAMID

34. Oblique Projection Objects with curves are best presented with the curves in the front face and drawn by compass or ordinate method.

Bracket The shape shown is arranged with the compass curves in the front face, depth lines at 45°. Compass radii for the rear face are the same as for the front face.

Circles on Cube Faces Circle on the front face may be drawn by compass. If the third axis is at 30°, compass arcs may be used in the ellipse, cavalier or full depth being used. Ellipses in the top face and end face, if foreshortening is used, are drawn by simple projection on centre lines and ordinates, as in the isometric diagrams.

Cylinder Where possible arrange the ends of the cylinder on the front and rear faces where the circles can be drawn by compass.

Bearing Block Simple arrangement of grouped solids drawn in the easiest position for quick mechanical representation.

Bearing Bracket Solid composed of cylinder and two prisms. Construct the 'boxes' for the three parts; the intersection of the sloping lines of the prism and cylinder is tangential.

Casting Draw the centre line and centres for the two bosses and draw the circles of the front plane. Draw the third axis depth lines, and construct the rear face circles.

Sphere on Base Draw the base, foreshorten if desired. The sphere is usually drawn on an extended radius to make the drawing more proportionate. The radius can be lengthened from R to R^1 by using 30° and 45° angles as shown in the diagram.

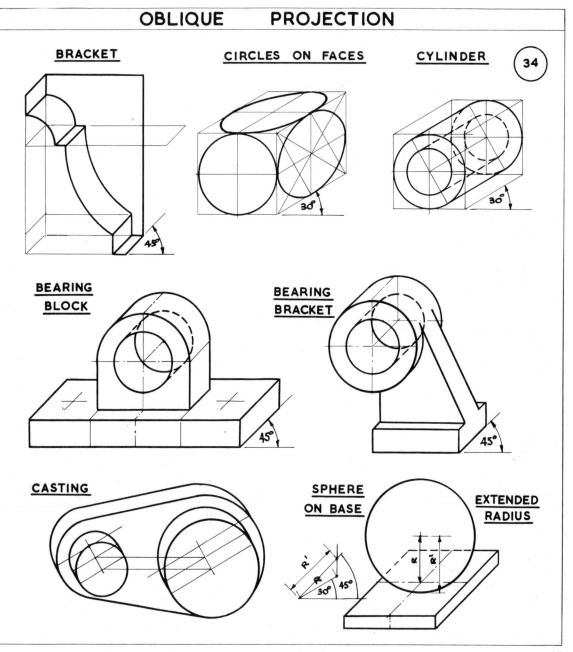

OBLIQUE PROJECTION

BRACKET

CIRCLES ON FACES

CYLINDER

34

BEARING BLOCK

BEARING BRACKET

CASTING

SPHERE ON BASE

EXTENDED RADIUS

35. Axonometric Pictorial Projection A pictorial method of drawing used by architects. A plan is drawn as a basis, turned through 45°, vertical lines are then drawn to give the height. Circles and arcs from the plan can be drawn conveniently by compasses as shown in the diagrams on the top face of the object, but circles appearing in the elevations are projected as ellipses in the pictorial view. Measurements for axonometric projections can be taken F.S. or to scale from the orthographic drawings and used direct on the axes or for ordinates. No measurements can be made on diagonals.

36. Perspective Drawing In this type of pictorial conical projection, projectors from points on the object converge at the eye point and pass through the vertical picture plane on which the image is drawn. The intersection of the picture plane and the horizontal ground plane gives the ground line—the horizon, on which lie the vanishing points, and is the same height from the ground line as the eye point is above the ground plane. The layout of the object, picture plane, eye point and eye line with the vanishing points may be drawn to a suitable scale as shown in the diagrams.

Join points 1 to 10 in the plan to the eye point, dropping perpendiculars at the points 1′ to 10′ on the picture plane. The heights of these lines on the picture plane may be obtained by measuring their true height on the line A B from the elevation, and joining to the appropriate vanishing point VP^1 or VP^2. This will give intersection points from which the perspective view can be drawn. Curves can be drawn by projecting points d, e, f, as in the diagram and drawing a fair curve through the points.

Note that the vanishing points VP^1 and VP^2 are obtained by drawing a line from the eye point parallel to the edges of the object and cutting the eye line.

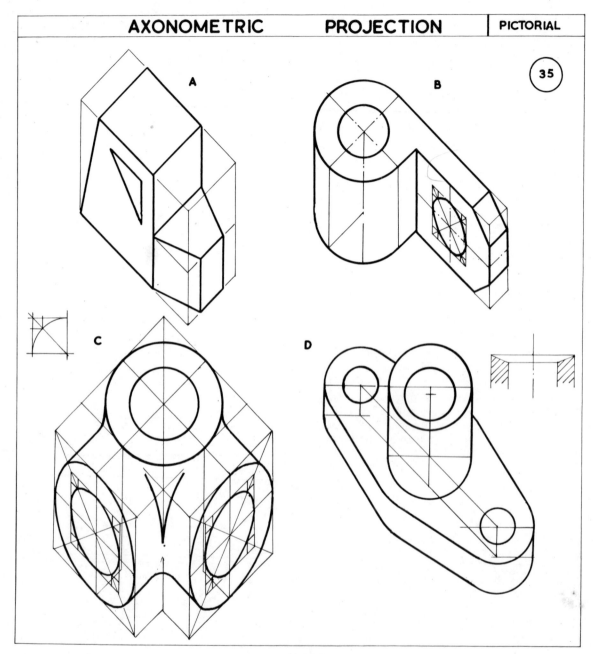

PERSPECTIVE DRAWING

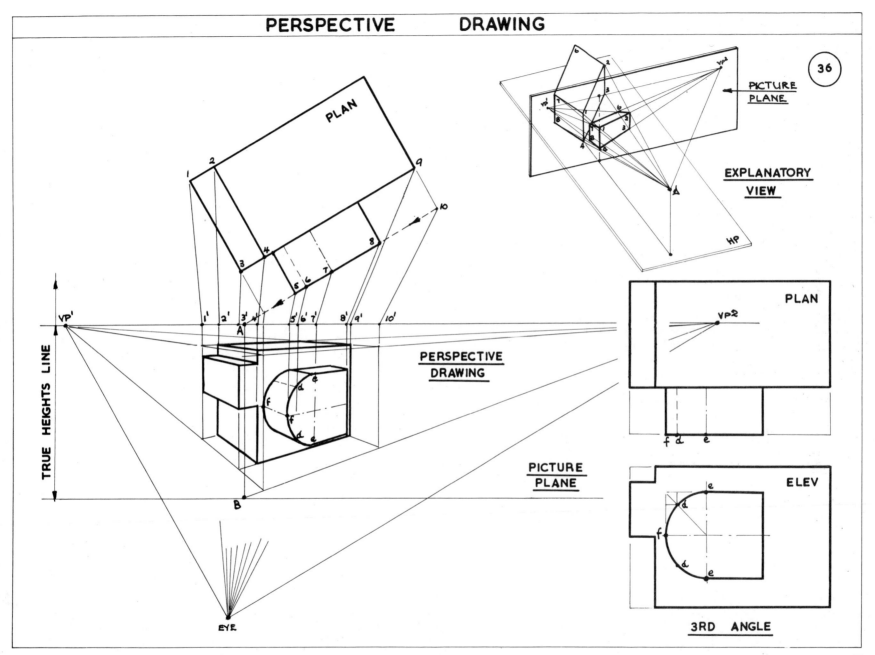

PLAN

EXPLANATORY
VIEW

PICTURE
PLANE

HP

36

VP²

PLAN

VP²

f d e

ELEV

e

d

f

d

e

3RD ANGLE

TRUE HEIGHTS LINE

VP¹

A

PERSPECTIVE
DRAWING

PICTURE
PLANE

B

EYE

d e

f

f

d e

17

37. Tall Objects Lines which are parallel to each other in the object, are drawn vanishing through the same VP; the picture plane may be any height, but the eye line is always at the same height above the base line as the eye point is above the ground plane.

Distant Vanishing Points To avoid quick diminishing of the planes of the object viewed, a large drawing board or table top may be employed with straight edges pivoted at the widely spaced VP points. This will result in less distortion.

Shadows in Perspective

Shadows may be caused by obstructions in the path of light from either the sun, in which case the rays are taken as being parallel, or from a lamp giving radial rays.

Sun Shadows The angle of the rays may be decided, 45° or 60° can be conveniently drawn by setsquare, and lines are projected from the upper corners of the object. Horizontal lines are then drawn from the lower appropriate points to intersect the former angle lines, and the shape of the shadow obtained by joining the points.

Lamp Shadows Lines are drawn from the point source of the lamp to the upper corners of the object, to intersect lines from the ground point of the source.

Exercises for Pictorial Drawing

38. Nine objects are shown suitable for drawing in isometric, oblique, axonometric and perspective. Where details have been omitted, discretion should be used.

39. Nine objects are shown to be used for drawing as above. One size only is given, the rest of the form should be estimated by proportion, giving practice in visualising the object.

18

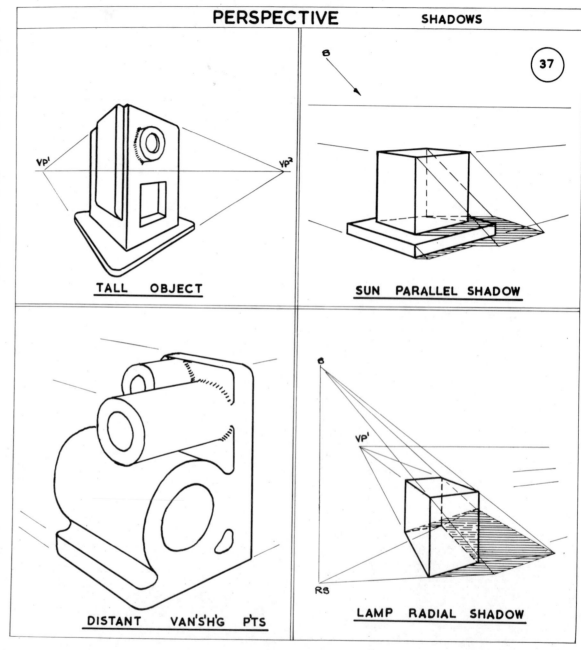

PERSPECTIVE   SHADOWS

37

TALL OBJECT

SUN PARALLEL SHADOW

DISTANT VAN'S'H'G P'TS

LAMP RADIAL SHADOW

EXERCISES FOR PICTORIAL DRAWING

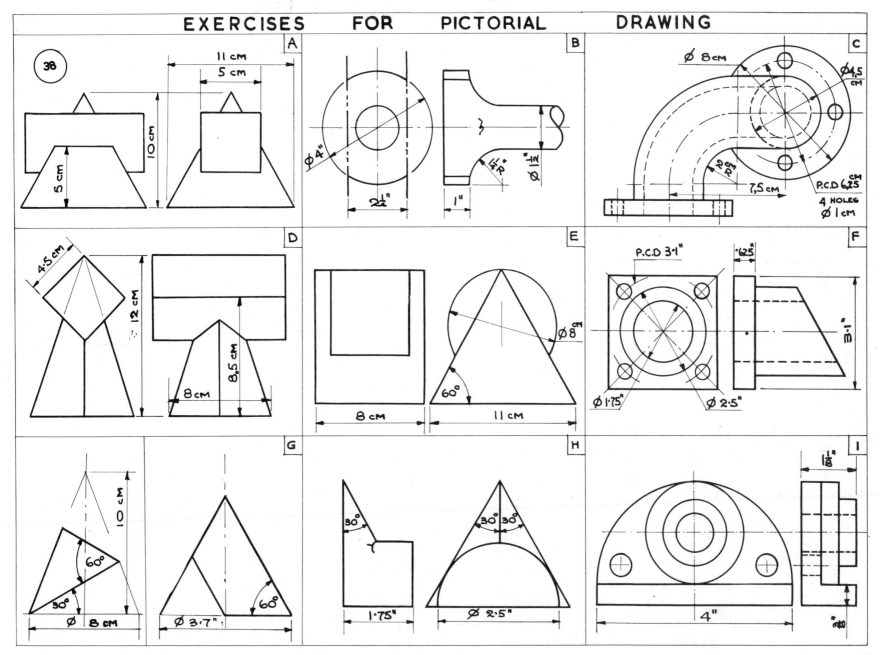

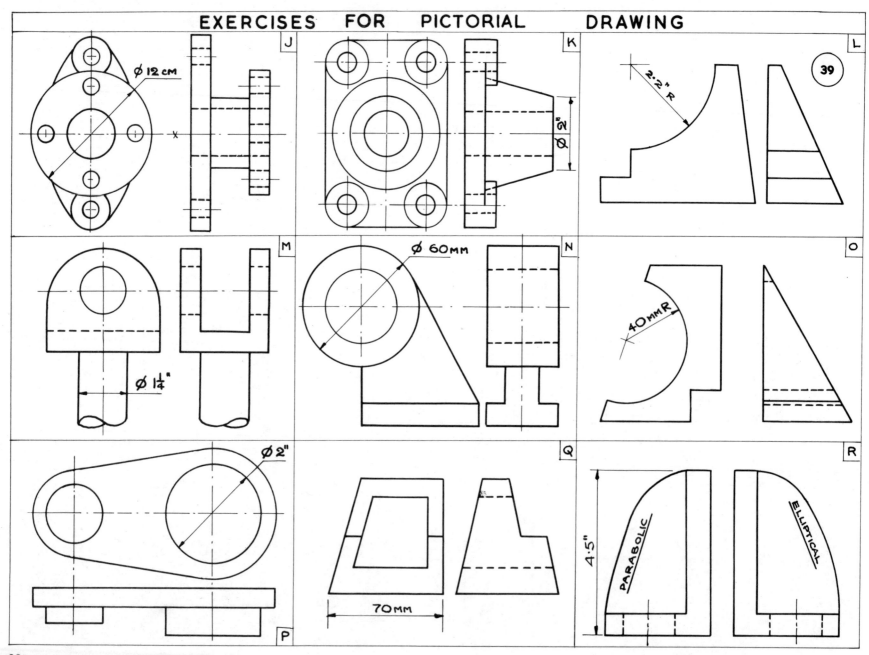

40. Isometric Drawing The diagram shows a cutaway isometric drawing of a Rotary Vane Pump, No. 385 in the Machine Drawing section, which has been used as a subject.

Other involved isometric drawings are given in Nos. 415 to 423.

Begin by drawing the base, working from a centre line. Extend the centre line, and draw the enclosing square of the front circular face. Construct the isometric ellipse by the four arc method shown in No. 29. Construct the ellipse for the rotor, which is offset. The blades' position can be found by ordinates. The shaft and oval pressure plate holding the seal in place are projected from a centre line.

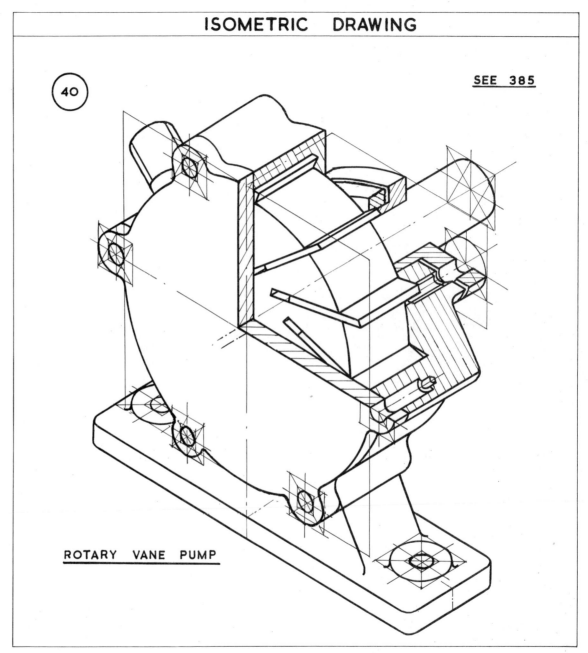

SEE 385

(40)

ROTARY VANE PUMP

21

CONICS

Ellipse

Parabola

Hyperbola

Evolutes

41. The Ellipse When a sectional plane cuts all the generators of a cone on one side of the apex, the section is an ellipse.

42. The Ratio of Eccentricity and position of directrix and focal points are obtained by bisecting the angle at V and drawing the circle which represents the focal sphere. Lines through the tangential points shown in the diagram show the positions. The ratio is that between the distance of the focal point from the vertex and the distance of the vertex from the directrix.

43. Ellipse Construction using a rectangle whose sides are in the ratio of eccentricity. The longer sides are distances from the directrix and the shorter from the focal point.

44. Parts of the Ellipse The diagram shows the parts of the ellipse and methods of drawing the normals and tangents. Given the major axis and focal points, the ellipse may be drawn by dividing the operative length of the major axis any number of times into lengths A and B and swinging arcs of A from one focal point and lengths of B from the other focal point. This satisfies the rule that an ellipse is the locus of a point which moves at a constant distance from two focal points.

Exercises

1. Draw a cone, height 6″ and base circle 4″, cut by a plane inclined at 45° to the H.P. $\frac{1}{2}$″ from the H.P. at a generator. Draw the derivations of the focus, vertex and directrix from the focal sphere as shown in the diagram.

2. A cone, 12 cm high with a base circle of 5 cm radius, is cut by an inclined plane which is at 30° to the H.P. The major axis of the section is 8 cm in length. Construct as in the diagrams, draw the focal sphere and obtain data to draw the section ellipse.

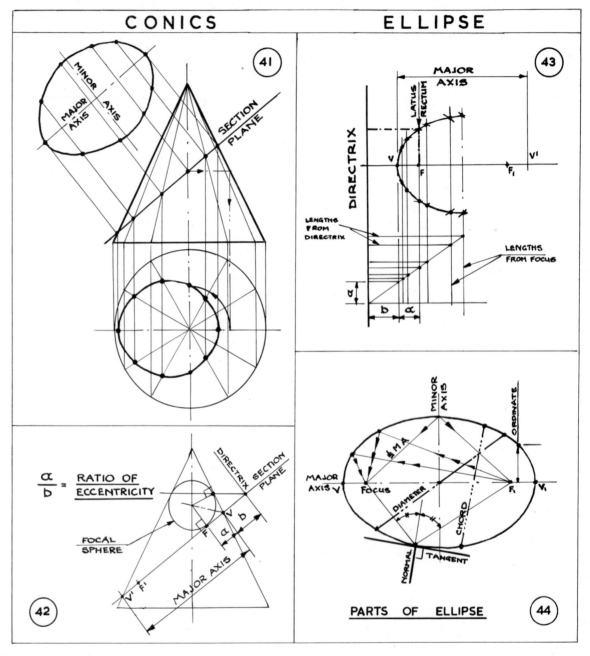

45. Auxiliary Circle Method of drawing an ellipse. Draw the two circles equal to the major and minor axes. Radials and projectors will give points on the ellipse.

46. Trammel Method Mark the half major and half minor axes on a strip of card and use this trammel to generate the curve, as in the diagram.

47. Enclosing Parallelogram.

48. Radial Interceptors These methods are self explanatory from the diagrams given.

Exercises

1. Draw an ellipse by the auxiliary circle method, major axis 5″, minor axis $2\frac{1}{4}$″.

2. Construct an ellipse by the trammel method. Major axis $6\frac{1}{2}$″, focal points 1″ from the vertices.

3. Construct an ellipse by the radial method in a rectangle 12 cm by 7 cm.

4. Construct by intersecting arcs from the foci, an ellipse whose major axis is 4·3″ long, and foci 2·7″ apart.

5. Given a focus $1\frac{1}{2}$″ from the directrix, construct an ellipse of eccentricity $\frac{3}{2}$.

6. A right cone height 4″ base circle 3″, is cut by a plane 45° to the H.P. Project the largest ellipse this plane can produce from the cone.

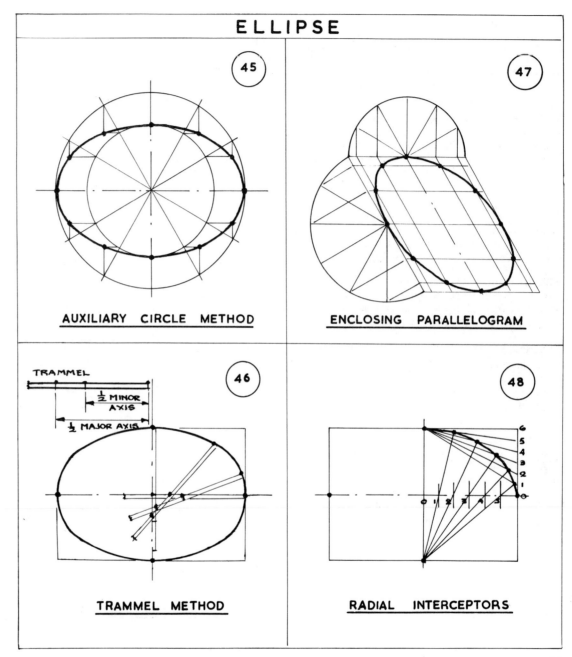

ELLIPSE

AUXILIARY CIRCLE METHOD

ENCLOSING PARALLELOGRAM

TRAMMEL METHOD

RADIAL INTERCEPTORS

3

49. Arc Methods of drawing approximate ellipses. First method. Draw the enclosing rectangle. Bisect the shorter side and join to the end of the minor axis. Join the corner of the rectangle to the other end of the minor axis. The intersection gives one point on the ellipse. Draw and bisect the two chords shown. This will give two centres for the four arcs of the ellipse.

50. The Second Method shows the enclosing rectangle with a diagonal drawn. Draw a quadrant with half the major axis as radius. Deduct the radius of the small construction circle from the diagonal. Bisection of the remaining portion of the diagonal gives the centres for the four arcs required.

51. The Isometric Method of drawing an approximate ellipse. Draw first the isometric parallelogram. Draw the two diagonals. Draw the centre lines through the intersection. Join apexes to mid points of sides and draw arcs from the centres shown in the diagram.

52. Section of Cylinder used to obtain an ellipse. Make diameter of cylinder equal to minor axis. Make length of sectional cut equal to major axis. Project as in the diagram and draw a curve through the intersections.

Exercise

Draw ellipses by each of these methods, 6″ × 4″ for 1 and 2; in an isometric square of 4″ for the isometric method; and using a cylinder 5″ high and 2½″ diam. cut to give an ellipse of 5″ major axis, in the last method.

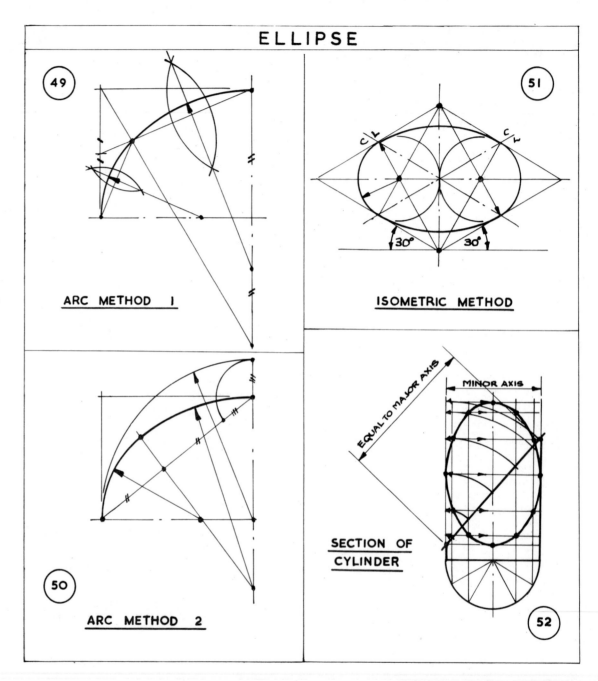

ELLIPSE

49

ARC METHOD 1

50

ARC METHOD 2

51

ISOMETRIC METHOD

SECTION OF CYLINDER

52

53. Ellipse Problems given the major axis and foci. Draw the major axis and mark foci. Bisect axis and draw perpendicular for minor axis. With radius half major axis and foci as centres, describe arcs cutting perpendicular to give minor axis length. Draw the ellipse by one of the methods shown previously.

54. Given Two Conjugate Diameters, draw the diameters as given and from P draw a perpendicular to diameter A and equal to half A. Join B to centre O, and draw circle, BO as diameter. Draw a line from P to pass through the centre of this circle to give x and y. The major axis lies on a line passing through xO, the minor axis passing through yO. Half the major axis is Py, half the minor axis is Px. Notice that the tangent of one conjugate diameter is parallel to that of the other conjugate diameter.

55. Given the Ellipse find the Axes. Draw any two parallel chords. Bisect these and draw a line to pass through the midpoints. Draw two other parallel chords, bisect and draw a similar line. Where these two diameters cross, draw a circle which will cut the ellipse in four places. Join to the centre, bisect the angles and draw in the axes.

56. Tangent to a Point on the Ellipse. Join the point to both foci. Bisect the outer angle and draw tangent through P. If the inner angle is bisected, the line through P is the normal, and the tangent can be drawn at right angles through P.

Tangent to a Point Outside the Ellipse. With radius equal to the major axis draw an arc from F. With centre P_1 and radius P_1–F_1 draw an arc to cut the previous arc. Join these intersections to F. Draw tangents to pass through the points where these lines cut the ellipse.

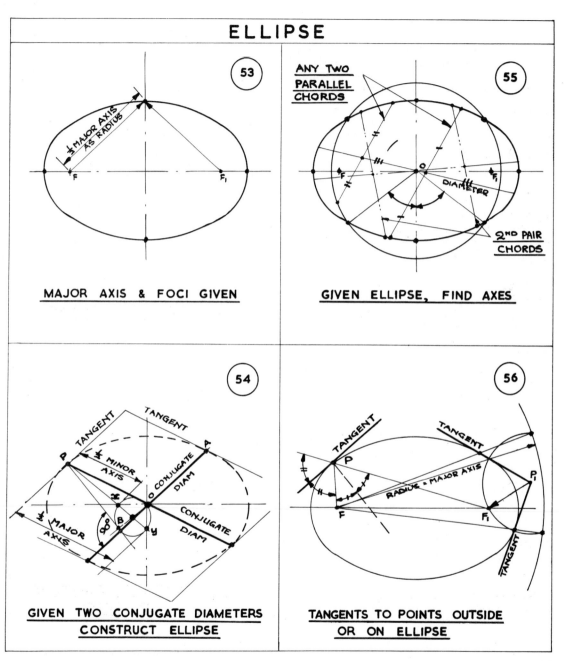

ELLIPSE

MAJOR AXIS & FOCI GIVEN

GIVEN ELLIPSE, FIND AXES

GIVEN TWO CONJUGATE DIAMETERS CONSTRUCT ELLIPSE

TANGENTS TO POINTS OUTSIDE OR ON ELLIPSE

57. Centre of Curvature on a point of the ellipse. Join P to the focus F. Draw the normal from P and where this crosses the axis erect a perpendicular to cut the line from F. Now erect a perpendicular to line P F to cut the normal at C; this is the point of curvature and a circle drawn on this centre will be tangential at point P.

58. Centre of Curvature at the vertex. Describe the arc centre F_1 and radius $V-F_1$. Extend the line by distance V F, and return a parallel to give C on the major axis.

59. Evolute of an Ellipse (1) Proceed by the method shown in the previous diagrams plotting the centres from points P_1, P_2, P_3. Join the centres of curvature in a fair curve, this is the evolute.

60. Evolute of an Ellipse (2) Where half the minor axis is less than the distance F O, then the two points of the evolute will fall outside the ellipse, otherwise the procedure for plotting the centres is the same.

Exercises

1. Draw an ellipse having axes of 6″ and 3″. Construct the evolute of the ellipse.

2. Construct an ellipse whose major axis is 5″ and foci 4″ apart. Construct the evolute.

3. Construct an ellipse 0·7 eccentricity, distance of focus from directrix 4 cm. Show the centre of curvature of a point on the ellipse 6 cm from a vertex.

4. Draw an ellipse which is enclosed by a rectangle of base 6″ and equal in area to a square $4\frac{1}{2}″ \times 4\frac{1}{2}″$.

5. Draw an ellipse. Data: major axis 15 cm, eccentricity $\frac{3}{4}$, vertex to focus 1·5 cm.

6. In the same diagram, from a focus $1\frac{1}{2}″$ from the directrix, draw an ellipse of eccentricity $\frac{1}{2}$.

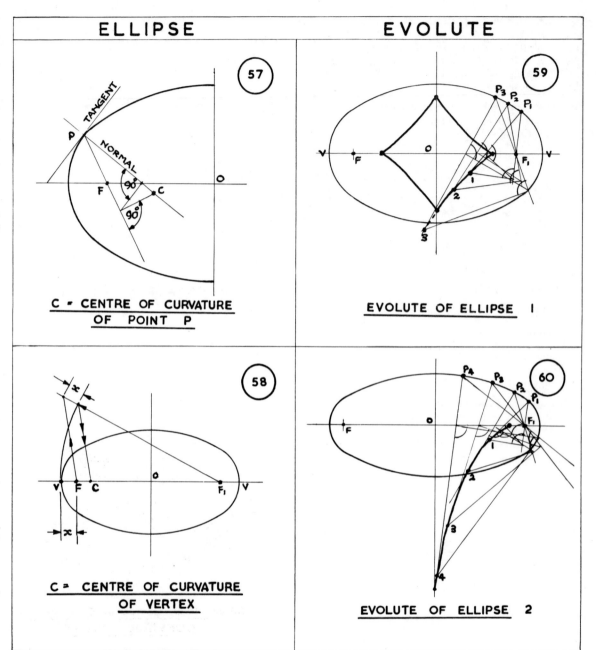

ELLIPSE   EVOLUTE

C = CENTRE OF CURVATURE OF POINT P

C = CENTRE OF CURVATURE OF VERTEX

EVOLUTE OF ELLIPSE 1

EVOLUTE OF ELLIPSE 2

61. The Parabola If a right cone is cut by a section plane parallel to a generator, the parabola results. Details of the method of projection including the true shape of the section are shown. The ratio of eccentricity is unity, i.e. the distance of the vertex from the directrix is the same as the distance of the focus from the vertex.

62. The Relations of the Directrix, the vertex and the focal point and method of derivation are given in the diagram.

63. The Parabola may also be constructed by drawing a centre line and directrix and plotting points of the curve by taking equal distances from the directrix and the focal point. A tangent at P may be drawn by joining P to F and drawing a perpendicular from the directrix to pass through P. Bisect the angle at P, this is the line of the tangent. The normal is drawn at right angles to the tangent. A parabolic reflector will emit a parallel beam of reflected light if the filament is on the focal point. The parabola is also the path of a falling body or a jet of water.

64. Approximate Methods of drawing the parabolic curve by radial intersectors in an enclosing rectangle are given.

65. Tangent to a Point outside the curve is obtained by first joining P to F and drawing a circle with FP as diameter. Draw an ordinate through the vertex and the tangents are drawn through the points of intersection and P. The actual point of contact is found by extending the tangent line until it intersects the directrix and joining this point to F. If a perpendicular to this line is drawn from F to cut the curve, then this is the actual point of tangency.

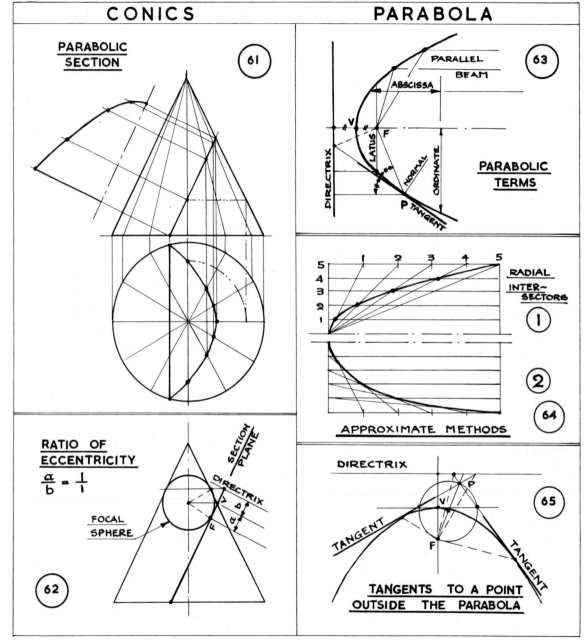

CONICS   PARABOLA

PARABOLIC SECTION   61

63   PARALLEL BEAM   ABSCISSA   V   F   DIRECTRIX   LATUS   NORMAL   ORDINATE   P TANGENT   PARABOLIC TERMS

RADIAL INTER~SECTORS   1   2   64   APPROXIMATE METHODS

RATIO OF ECCENTRICITY $\frac{a}{b} = \frac{1}{1}$ FOCAL SPHERE   SECTION PLANE   DIRECTRIX   62

DIRECTRIX   65   TANGENT   V   P   F   TANGENT   TANGENTS TO A POINT OUTSIDE THE PARABOLA

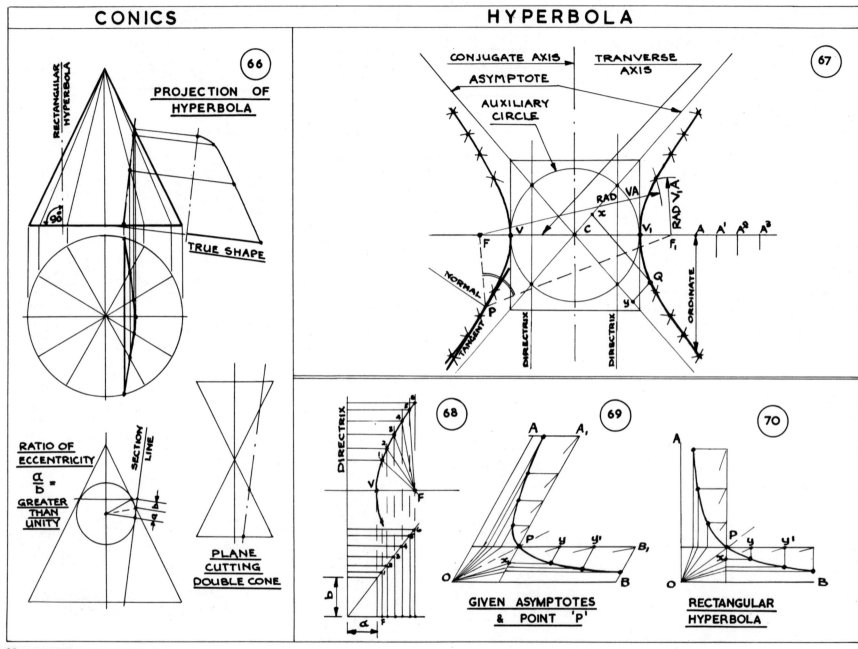

CONICS

66 PROJECTION OF HYPERBOLA

RECTANGULAR HYPERBOLA

TRUE SHAPE

RATIO OF ECCENTRICITY
$$\frac{a}{b} = \text{GREATER THAN UNITY}$$

SECTION LINE

PLANE CUTTING DOUBLE CONE

HYPERBOLA

67

CONJUGATE AXIS — TRANVERSE AXIS

ASYMPTOTE

AUXILIARY CIRCLE

RAD VA

RAD V,A

NORMAL

TANGENT

DIRECTRIX   DIRECTRIX

ORDINATE

A A' A^2 A^3

68 DIRECTRIX

b

a

69

GIVEN ASYMPTOTES & POINT 'P'

70

RECTANGULAR HYPERBOLA

66. The Hyperbola If a section plane cuts the double cone on one side of the axis, the section is a hyperbola. The diagrams show the projection of the true shape of the cut. Details of the focal point, the vertex and the directrix are obtained from intersections of the focal sphere and the section plane and are shown in the lower diagram. The ratio of eccentricity is greater than 1 (unity).

67. Method of Drawing the Hyperbola Draw the transverse axis and the conjugate axis. Draw the auxiliary circle. Mark in the two focal points and the two vertices. Draw the two directrix. Draw the two asymptotes through their intersection with the auxiliary circle. The curve is drawn through points obtained by intersecting arcs drawn from the foci. The radii of the arcs is obtained by first marking off suitable points A, A^1, A^2, A^3, and using radius VA from one focal point and radius V_1A from the other focal point. The four intersections give points on the hyperbolic curve. Other points may be found by using radii obtained from A^1 etc.

68. Given the Ratio of Eccentricity the hyperbola may be drawn by first drawing the directrix and the transverse axis and stepping off the vertex and the focal point in the ratio given (3:2 in this case). The rectangular diagram drawn to the same proportion gives values of a and b to be used in obtaining points for the curve, a lengths being perpendicular from the directrix and b lengths as intersecting radii from the focal point F_1, F_2, F_3, etc.

69. Given the Asymptotes and a Point on the curve. Draw the two asymptotes A O and O B and the point P. Draw lines A_1 and B_1 parallel to AO and OB and passing through P. Draw a number of suitable radials from O. Where these cut A_1 and B_1, construct parallelograms. Compare the points x y and point Q in the upper drawing. The curve is drawn through the corners of the similar parallelograms.

70. Rectangular Hyperbola The construction is similar to that used in the previous diagram.

71. The Parabolic Evolute The loci of the centres of curvature of the parabola is drawn in the same way as that of the ellipse already shown. Take suitable points on the parabola 1, 2, 3, 4 and draw normals from these points. Draw also a line from each point through the focal point. From the point where each normal crosses the axis draw a perpendicular to cut the focal chord. From this point draw a perpendicular to the focal chord to cut the normal. This second point on each normal is the centre of curvature for the original point on the parabola. In the case of the parabola, the C of C for a point at the vertex is the focal point itself. The complete method is shown in the diagram.

72. The Hyperbolic Evolute Points on the curvature are obtained by the same method as above by drawing the normal and the focal chord, then the perpendiculars to arrive at the final point on the normal. The evolute is drawn through the points. In the case of the hyperbola, the centre of curvature for a point at the vertex is shown in the small diagram, and it should be noticed that the distances V_1–F_1 and F_1–O are in the same ratio as the ratio of eccentricity of the hyperbola.

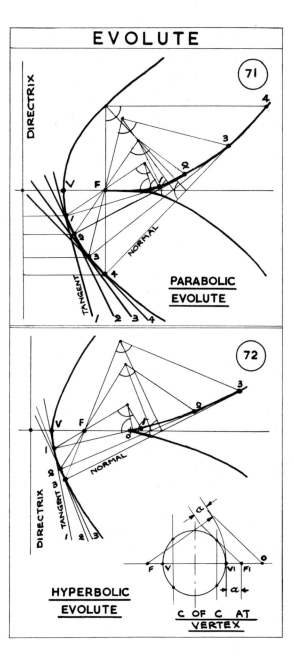

EVOLUTE

PARABOLIC EVOLUTE

HYPERBOLIC EVOLUTE

C OF C AT VERTEX

73. Hyperbolic Section given the cone and section plane. Draw the double cone and section plane. Draw the focal spheres, the centres of which are found at the intersection of the axis and a bisector of the angle of the cone and the section plane. The focal points and vertices can now be projected to the centre line. Project the directrix lines from the planes which pass through the tangent points of the focal spheres as shown. Draw the auxiliary circle, and where the directrix lines intersect this circle, draw the asymptotes to pass through centre C. The hyperbolic curves can be drawn using the method shown on a previous page, dealing with the hyperbola.

Exercises

1. Draw the hyperbolic curve at the section shown in the sketch.

2. Draw the rectangular hyperbola given by a plane which cuts a cone of 5″ height and 4″ diam. giving a chord line in the base circle of $3\frac{5}{8}''$.

3. Draw the hyperbola from the following data. Focal length $1\frac{3}{4}''$, eccentricity $\frac{4}{3}$, base line $4\frac{1}{2}''$.

4. Draw the evolute of the hyperbola of the previous question.

5. Given two asymptotes at 60° and a point lying on the bisector of the angle $1\frac{3}{8}''$ from the intersection, construct the hyperbolic curve.

6. Construct a hyperbolic curve when two asymptotes are at right angles and a point on the curve is 1″ from the junction.

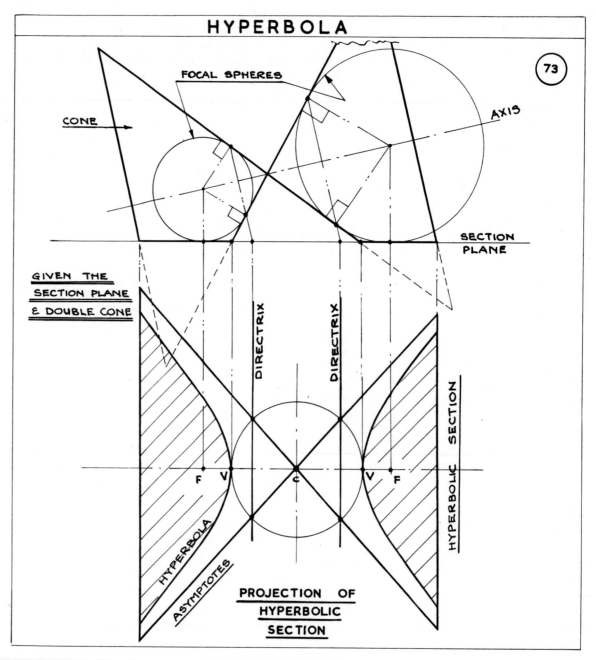

74. Elliptical Section of Right Cone, given the cone and the section plane. Draw the given elevation and the plane. Project the vertices V V, this gives the major axis of the ellipse. Draw the focal sphere; the centre can be obtained by drawing the axis line of the cone and cutting this line with a bisector of the angle at A. Project the centre of the focal sphere to the major axis giving the focal point F. The second focus can be also fixed.

The minor axis found by intersecting arcs form the focal points radius V O or half the major axis. The ellipse can now be completed by one of the methods given on an earlier page.

The directrix line is projected from the plane which passes through the two tangent points of the focal sphere and the generators of the cone, as shown in the diagram.

75. Parabolic Section of a Right Cone, given the cone and the section plane. Draw the cone and plane, draw the focal sphere, and project the centre to give F the focal point on the centre line. Obtain the directrix line by projecting the line which passes through the tangent points shown of the focal sphere. The parabolic curve can now be drawn by the method shown on an earlier page.

76. Exercises

Ellipses

1. Construct the ellipse given by the section plane in the first sketch shown. (Turn the cone until the cutting plane is horizontal, the question is then similar to that shown.)

2. Draw the ellipse required by the second diagram: the major axis will be 9 cm.

Parabolas

3. Draw the parabola which would be given by the cutting plane to the requirements of the third sketch.

4. The fourth sketch shows a cutting plane cutting a cone parallel to a generator and passing through the centre point of the centre line. Construct the parabola. Construct the hyperbolic curves shown when the double cone is cut as shown.

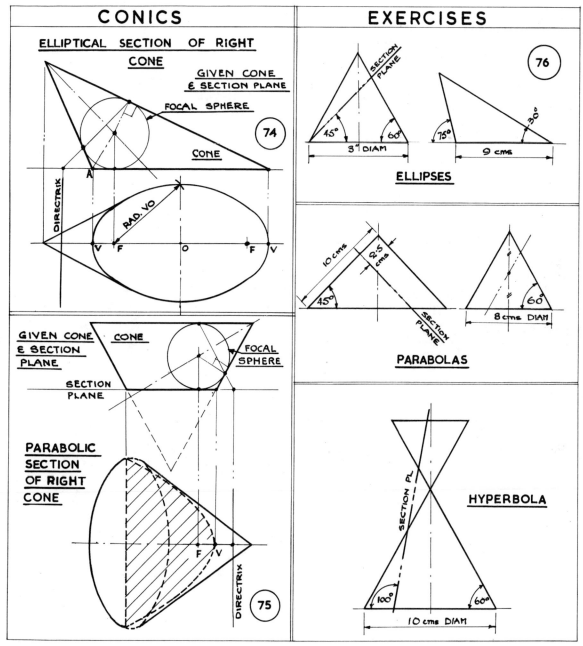

76A. Conic Problems

I. If a focal chord is drawn through the focal point of an elliptical or parabolic curve as at AB in the diagram, tangents to the curve at A and B will be at right angles to each other and intersect in the same point on the directrix.

II. Given the asymptotes and the curve of the rectangular hyperbola, find the focal point. Refer to No. 67 for construction, draw the asymptotes, auxiliary circle, transverse axis, conjugate axis and the directrix. From the intersection of circle, asymptote and directrix, draw a line at right angles to the asymptote to cut the transverse axis in the required focal point. (The hyperbolic curve used in this example is that used in No. 70.)

III. A similar example is shown using the hyperbolic curve shown in No. 69. Draw the curve first and the asymptotes, proceed then with the construction as above.

IV. Given the asymptotes and the auxiliary circle, find the focal point and draw the hyperbola. Draw the transverse axis, auxiliary circle, conjugate axis, asymptotes and directrix. From the intersection of a circle, asymptote and directrix draw a line perpendicular to the asymptote cutting the transverse axis in the focal point. Draw the hyperbolic curve by the method of intersecting arcs shown in No. 67.

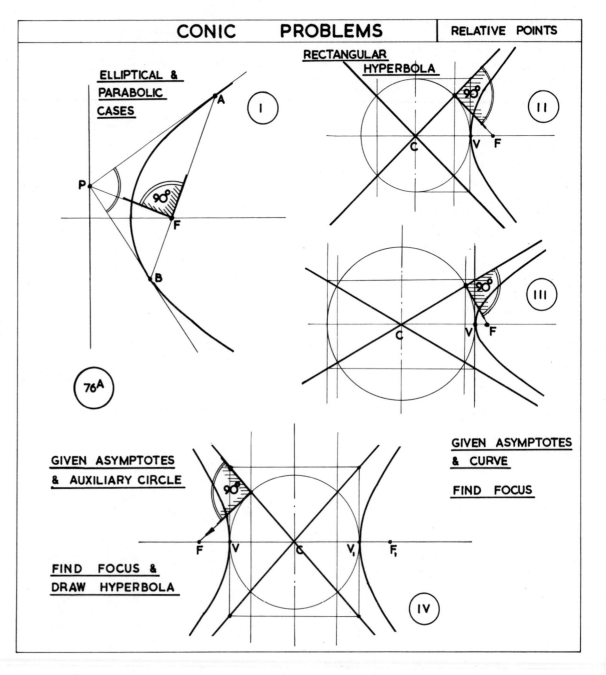

CYCLOIDS

Trochoids

Evolutes

77. Cycloid The locus of a point on the circumference of a circle rolling without slipping on a straight base line. Draw the generating circle and the line. Draw the twelve generating points on the circle, mark off twelve parts on the line to equal the circumference. Project lines from the circle to intersect the perpendiculars shown. From the twelve centres strike arcs, radius as circle, to give the points on the cycloid. The process may be likened to the positions of a wheel spoke through one revolution.

78. Epicycloid and Hypocycloid If the circle rolls round the outside of a base circle, the plot of the point yields an epicycloid. When the circle rolls inside the base circle the locus is the hypocycloid.

Draw the base circle and the generating circles. Divide the generating circles by twelve and draw arcs from O. Step off on the base circle twelve divisions to equal the circumference of the generating circles, and draw radials from O. Obtain the centres of the rolling circles, and strike arcs, radius R^1 and R^2 to cut the arcs from O. Draw the curves through these points.

The length of the arc on the base circle is also given by using the formula to give the subtended angle at O. Two special cases are shown: (*a*) When the radius of the rolling circle is half that of the base circle the hypocycloid is a straight line. (*b*) The rolling circle may be larger than the base circle and thus have internal contact.

79. Centres of Curvature for the cycloids are shown. Given point P, find its centre and draw the generating circle. Find the point of contact with the base line. The line drawn through P and this point is the normal. Construct the tangent at P. Draw the enclosing rectangle, join the circle centre to O. The Centre of Curvature is at the intersection C. The evolute is the locus of such centres, and is a curve identical to the cycloid itself.

80. The Centres of Curvature and the evolutes for the epicycloid and the hypocycloid are shown. Proceed as before. From P, find the centre of, and draw the generating circle. Join P to the contact point at the base circle, giving the normal. Radials from O to the centre of the circle and two points on its circumference, indicate a smaller generating circle which forms the basis for another epicycloid which is the evolute of the original epicycloid.

The Centre of Curvature for the hypocycloid is obtained by drawing the generating circle for P_1; drawing the normal and tangent. The intersection of the normal and the centre line from O indicate the distant centre.

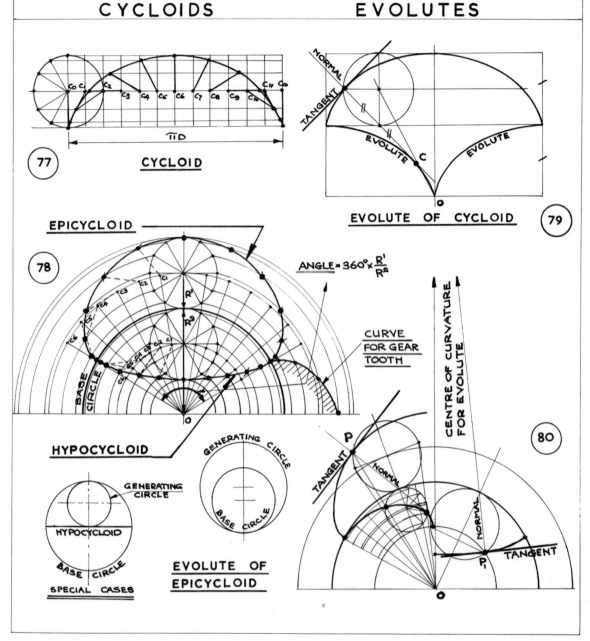

81. Inferior Trochoid The locus of a point which lies inside the generating circle. The construction layout is similar to that of the cycloid; notice that the 'spokes' will lie in similar positions, but the radius is that of the generating circle.

82. Superior Trochoid If the point lies outside the generating circle (on the extension of the 'spoke'), the locus is a superior trochoid.

83. Inferior Epitrochoid The locus of a point lying inside the generating circle which is rolling outside the base circle. Half the curve is shown and the construction is similar to that of the epicycloid.

84. Superior Epitrochoid The locus of a point lying outside the generating circle which is rolling outside the base circle.

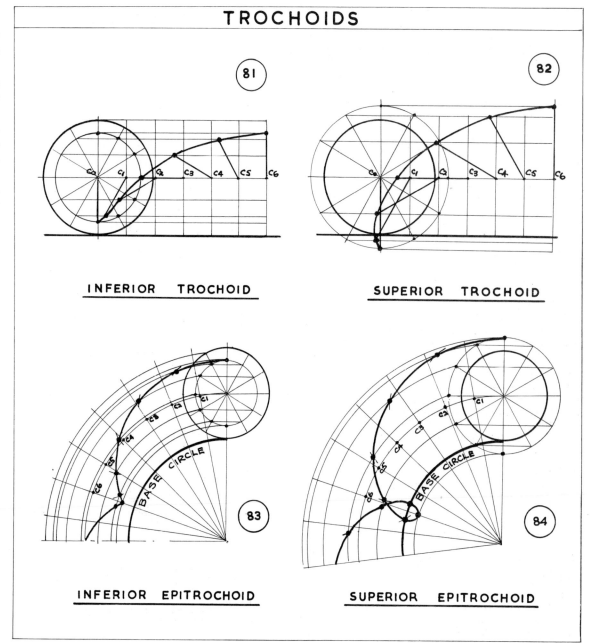

TROCHOIDS

INFERIOR TROCHOID

SUPERIOR TROCHOID

INFERIOR EPITROCHOID

SUPERIOR EPITROCHOID

85. Superior Hypotrochoid Locus of a point lying outside the generating circle which is rolling inside the base circle. Project as in the previous examples.

86. Inferior Hypotrochoid Locus of a point lying inside the generating circle which is rolling round the inside of the base circle.

87. Centres of Curvature for the hypotrochoids are shown. Given the point P find the centre of and draw the generating circle at that position. Join the centre to O which is the centre of the base circle. Join P and the contact point, to obtain the normal. The tangent is drawn at right angles to the normal and through P. Draw the right-angled triangle having P and contact point as base. Join as shown to the centre of the base circle. The centre of curvature is shown at the indicated intersection. Draw the evolute through several such points obtained by projections of P along the given epitrochoid. In the case of the inferior hypotrochoid the centre of curvature lies distant and is indicated by the diagram.

Exercises

Trochoids

1. Draw (*a*) an inferior trochoid, (*b*) a superior trochoid using a base circle of $2\frac{1}{2}$″ diam. and a point $\frac{3}{8}$″ inside or outside the radius as required. Draw only half of the complete curve.

2. Construct (*a*) an inferior epitrochoid, (*b*) a superior epitrochoid using a circle of 50 mm diam. as rolling circle, with a point 5 mm from the end of the diameter. The circles should roll round the outside of a 120 mm diameter base circle.

3. Construct the superior hypotrochoid shown in the diagram. Use a rolling circle of 2″ diam. turning inside a base circle of 3″ radius. The point should be $\frac{3}{8}$″ outside the moving circle.

4. Draw the inferior hypotrochoid shown, using a rolling circle of 50 mm diam. rolling inside a base circle of 10 cm radius.
   Draw the Centres of Curvature for the two last cases.

38

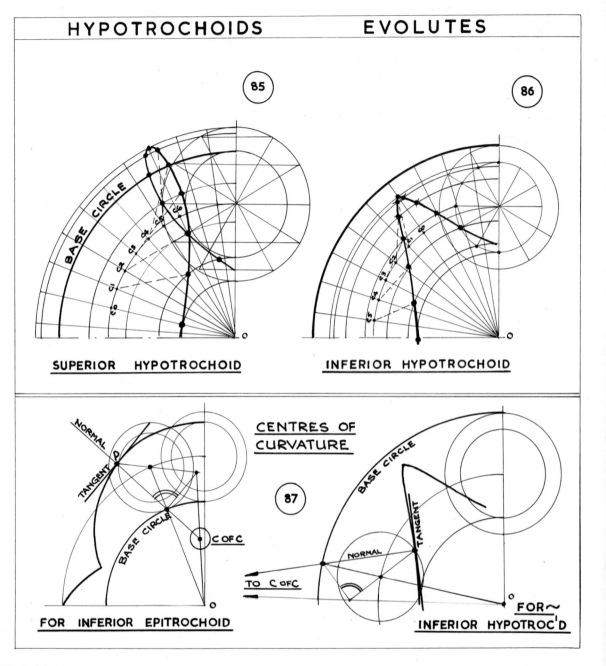

HYPOTROCHOIDS   EVOLUTES

SUPERIOR HYPOTROCHOID   INFERIOR HYPOTROCHOID

CENTRES OF CURVATURE

FOR INFERIOR EPITROCHOID   FOR~ INFERIOR HYPOTROC'D

88. Exercises

Cycloids

1. Construct a cycloidal curve for a point on a wheel $2\frac{1}{2}''$ diam. which rolls without slipping along a horizontal plane.

2. A circle 2″ diam. rolls without slipping round the perimeter of a 5″ diam. circle. Draw and name the locus of a point P on the circumference of the rolling circle.

3. Another similar circle rolls on the inside of the base circle of the previous question. Draw the locus of a point P beginning from a common point with the previous locus, the circle moving in the opposite direction. Show how the curves are used in gearing.

4. Construct the evolute of the curve drawn in Question 1.

5. Construct the Centre of Curvature for one position of a point P on an epicycloidal curve given when a circle of 4 cm diam. rolls on a base circle of 10 cm diam.

(a) Draw the three loci of points P, P₁, P₂ of the circle which rolls without slipping on the contour shown.

(b) Plot the loci of the point P on the circle which rolls without slipping inside the semi-circle shown. Draw only half the plot.

(c) Plot the loci of the three points of the circle shown in the diagram, which rolls without slipping on the given contour.

(d) A square rolls without slipping inside the circle as shown. Plot the loci of points P and P₁ for one circuit.

(e) A circle rolls round an ellipse as shown. Plot the locus of the three points of the circle as far as B.

(f) A triangle moves round the inside of the given circle without slipping as indicated in the diagram. Plot the loci of points P and P₁ of the triangle.

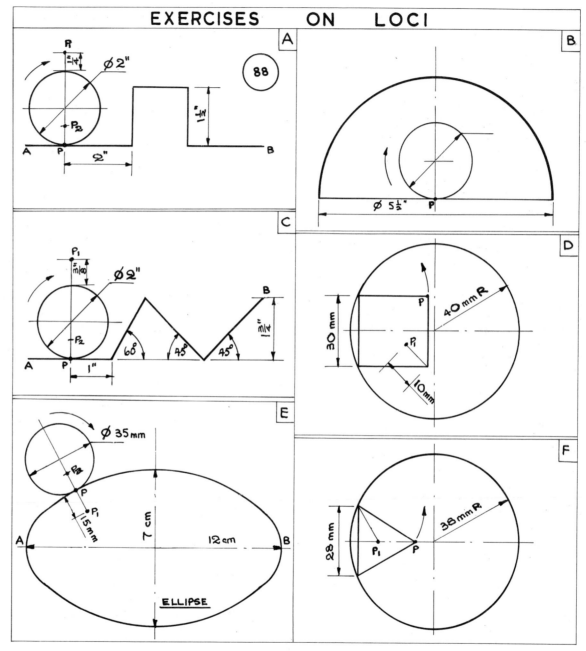

EXERCISES ON LOCI

39

88A. Cycloidal Problems.

A. A circle rolls round an ellipse without slipping. Plot the locus of point P, starting point as shown, for one revolution.

B. A circle rolls without slipping round an ellipse. Plot the locus of Q from its shown starting point.

C. A circle rolls round an ellipse without slipping. Plot the locus of R from the position shown.

D. A circle rolls inside the ellipse without slipping. Plot the locus of T for one revolution.

E. A circle rolls round the inside of an ellipse without slipping. Plot the locus of S for one revolution.

Draw the ellipse, major axis 5″, minor axis $3\frac{1}{2}$″; base circle A, B, C, 2″ diam.; base circle D, $1\frac{3}{8}$″ diam.; base circle E, $1\frac{1}{4}$″ diam.; other dimensions left to discretion.

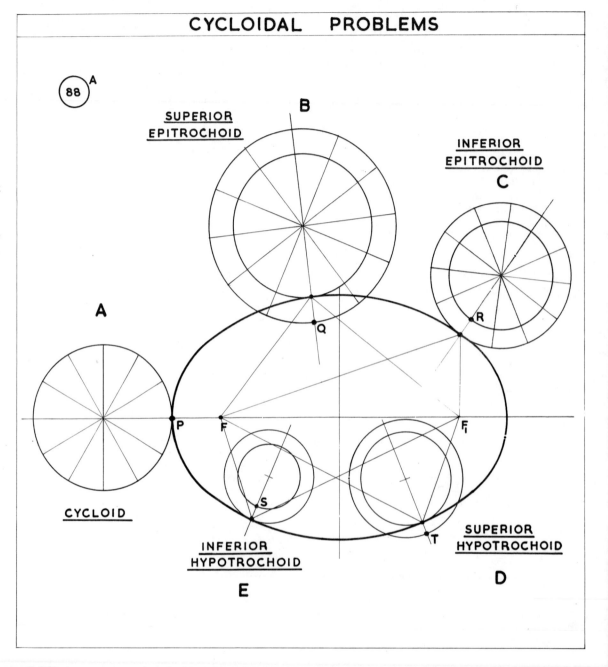

CYCLOIDAL PROBLEMS

88 A

SUPERIOR
EPITROCHOID

INFERIOR
EPITROCHOID
C

A

B

R

Q

CYCLOID

P F

F₁

S

T

INFERIOR
HYPOTROCHOID

SUPERIOR
HYPOTROCHOID

E

D

CAMS

Followers

Uniform Velocity

Simple Harmonic Motion

Uniform Acceleration
and Retardation

Radial Plate Cams

Cylindrical Cams

89. Plate Cams. If a disc is shaped to a suitable contour and mounted on a shaft it can be used to lift a follower and so a lever or a valve when the shaft turns. Followers may be knife ended, roller, flat or angled foot. They may be in line with the camshaft centre or offset. Knife followers can follow most contours but wear rapidly. Roller and flat followers cannot follow hollow contours and the cam profile must therefore be such that no bridging occurs. Examples of simple cam assemblies are given.

90. The Design of a radial plate cam is shown. First the performance graph is drawn showing the lift, the rest or dwell, the fall and the lower dwell. The graph may be in terms of degrees of rotation or in time, which would be calculated on the basis of one or more r.p.m. The points of the graph are now projected to the centre line of the disc which is marked out in degrees, and brought to the corresponding radial. These points are joined to give the contour of the cam.

All but simple symmetrical cams can only turn either in clockwise or anticlockwise rotation to give the desired performance, and this must be watched in plotting.

91. Uniform Velocity This term is applied to lifts or falls which follow a straight line on the graph; in practice the apex is rounded to avoid bounce of the follower. A cam is shown which would give uniform velocity rise and fall of its follower.

92. Simple Harmonic Motion This is the sine curve, and is constructed by projecting points from the half-circle to the graph lines. When projected to the cam, a circle results, see cam and flat follower at the top diagram.

93. Uniform Acceleration and Retardation This curve is drawn by fixing apex, base and midpoints, and in the four quarters thus made, drawing six radials and six verticals. The curve is drawn through the intersections. The shape of the finished cam is shown.

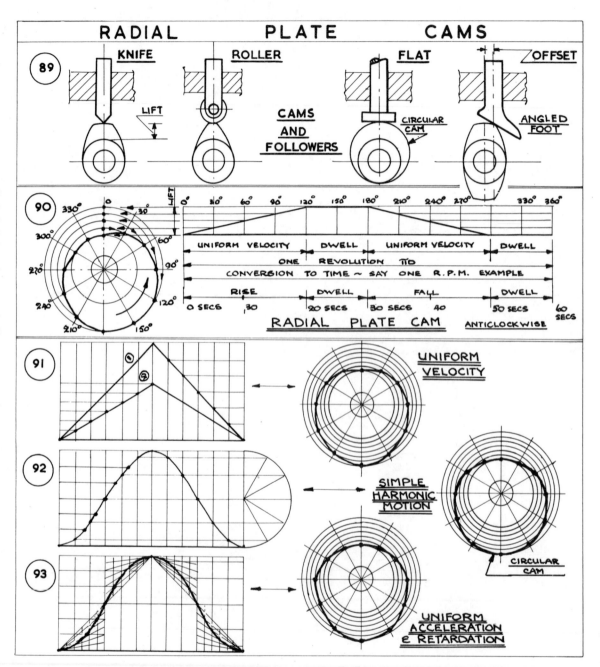

42

94. Radial Plate Cam Design a cam to fulfil the conditions in the given data. Draw the graph lines showing a 1″ lift and twenty-four 15° vertical divisions. Draw the straight line U.V. lift ½″ for 120°. Plot the S.H.M. curve on the next 90°, note the small semi-circle for the plot. A straight line horizontal gives the dwell or rest to 270°, and a straight sloping line to 360° gives the final U.V. fall back to the base line. Draw the circle for the cam and project the points on the graph to the centre line. Swing each point to its own radial and the intersections give the outline of the cam. This cam is suitable only for a knife follower as a roller or a flat could not follow the indentation.

95. Radial Plate Cam When a roller ended follower is used, the lift is measured from its centre, and the cam line is drawn tangential to the roller circles after plotting. Draw the graph to satisfy the given data, ⅞″ lift and 24 positions. Construct a Simple Harmonic Motion curve to lift ⅞″ in 180°, high dwell for 30° shown by the straight line, Uniform Acceleration and Retardation curve, to fall ⅞″ to the base line. The heights of the positions on the graph are projected to the vertical centre line of the circle and each swung to its own radial. Roller circles are now drawn on each centre, and the cam profile is obtained by drawing it tangential to the circles. This profile is also suitable for a flat ended follower as there are no concave curves.

Quick models may be made of plate cams by cutting in balsa- or plywood and mounting the shape on a pin, testing the performance against a plywood follower.

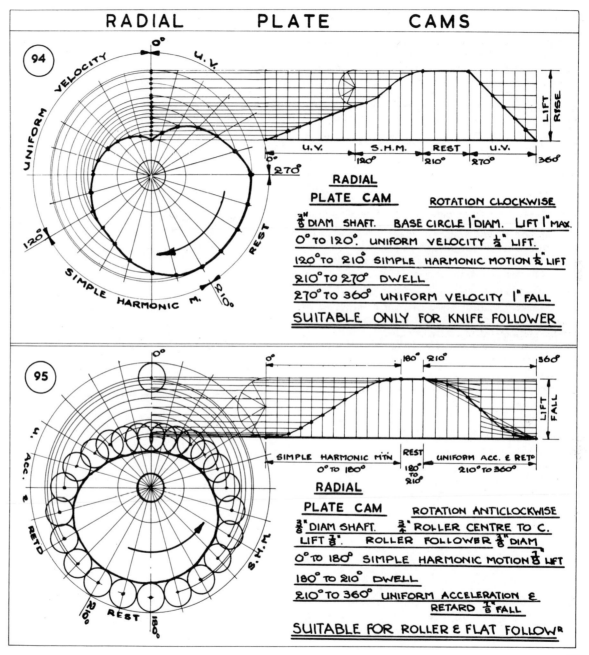

RADIAL PLATE CAMS

94

RADIAL PLATE CAM   ROTATION CLOCKWISE
⅜″ DIAM SHAFT.   BASE CIRCLE 1″ DIAM.   LIFT 1″ MAX.
0° TO 120° UNIFORM VELOCITY ½″ LIFT.
120° TO 210° SIMPLE HARMONIC MOTION ½″ LIFT
210° TO 270° DWELL
270° TO 360° UNIFORM VELOCITY 1″ FALL
SUITABLE ONLY FOR KNIFE FOLLOWER

95

RADIAL PLATE CAM   ROTATION ANTICLOCKWISE
⅜″ DIAM SHAFT.   ¾″ ROLLER CENTRE TO C.
LIFT ⅞″.   ROLLER FOLLOWER ⅜″ DIAM
0° TO 180° SIMPLE HARMONIC MOTION ⅞″ LIFT
180° TO 210° DWELL
210° TO 360° UNIFORM ACCELERATION & RETARD ⅞″ FALL
SUITABLE FOR ROLLER & FLAT FOLLOWER

96. Radial Plate Cam, Offset In this example the line of the follower is offset to the right by $\frac{5}{8}''$. Draw the cam circle, draw in the line of the offset and draw the first low position of the roller with centre on this line and tangential to the centre circle. Project the centre of the roller to give the base line of the graph. Draw the graph layout to give 1″ lift and 24 divisions in 360°. Draw the Uniform Acceleration and Retard curve for 120° and 1″ lift; straight line dwell for 30°, and a sloping straight line to 225° for the Uniform Velocity section, fall $\frac{1}{4}''$; and a Simple Harmonic curve to 360° falling $\frac{3}{4}''$ to the base line. Before being able to project the points to the circle, it is necessary to draw the tangent lines which are generated at the centre circle. The points from the graph are now projected as in the diagram and the roller circles drawn. The cam profile is drawn tangential to these.

97. Cam and Radial Arm The diagram shows a cam which lifts a radial arm instead of a vertical follower. Notice that the roller on the end of the arm moves forward as it passes through the centre of its travel.

98. Cylindrical Cam If a groove is machined in a cylinder, a suitably shaped radial arm can be arranged to follow the groove. A development is made of the cylinder and the motion plotted centre line first.

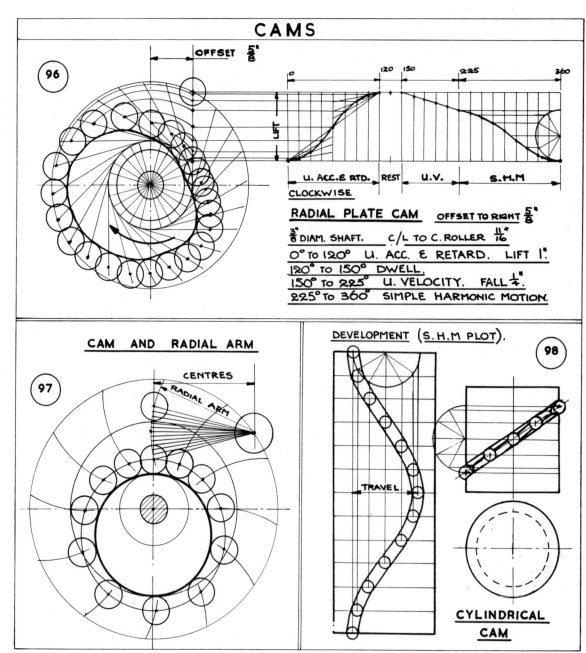

CAMS

96

OFFSET $\frac{5}{8}''$

LIFT

0   120 150   225   360

U. ACC.& RTD. | REST | U.V. | S.H.M

CLOCKWISE

RADIAL PLATE CAM   OFFSET TO RIGHT $\frac{5}{8}''$

$\frac{3}{8}''$ DIAM. SHAFT.   C/L TO C. ROLLER $\frac{11}{16}''$

0° TO 120° U. ACC. & RETARD. LIFT 1″.

120° TO 150° DWELL.

150° TO 225° U. VELOCITY. FALL $\frac{1}{4}''$.

225° TO 360° SIMPLE HARMONIC MOTION.

CAM AND RADIAL ARM

97

CENTRES

RADIAL ARM

DEVELOPMENT (S.H.M PLOT).

98

TRAVEL

CYLINDRICAL CAM

Exercises

99. Cams

1A. Design a plate cam to the following data: $\frac{1}{2}''$ diam. shaft, $1\frac{1}{4}''$ minimum diam., $1''$ lift. Clockwise rotation, knife edge follower. Performance: Uniform Velocity 180°, 30° dwell, Uniform Acceleration and Retardation fall 150°.

B. The sketch shows the requirements for a radial cam. Draw the cam to give a rise of S.H.M., short dwell, and fall of U.V. to the required time factor. (Five r.p.m. gives 12 sec/rev., i.e. the cam rotates through 360° in 12 sec = 30° per sec.)

C. From the cam given, draw its performance graph.

D. Draw the double lift cam shown, and draw its displacement graph.

E. The sketch shows the details of a cylindrical cam. Draw out the development of the cam surface, and project the required groove thereon, S.H.M.

F. A simple quick return cam operating a plunger is shown. Draw out the performance graph. State the form.

2. Draw a radial cam to give the following performance. Clockwise, 0°–120° S.H.M. $1\frac{3}{4}''$ lift, 120°–150° dwell, $1''$, 240°–360° U.V. fall $\frac{3}{4}''$. Shaft $\frac{1}{2}''$ diam., minimum diam. $1\frac{1}{4}''$.

3. Radial cam turns at 10 r.p.m. Show its performance graph to give a lift of 25 mm U.V., a 1 sec dwell, and a fall of 25 mm S.H.M. Flat ended follower, 6 mm offset.

4. Design a radial cam to work a radial arm, 2″ long, external diameter of cam 3″, minimum diam. $1\frac{1}{2}''$. S.H.M. rise and fall with 10° dwell and rest for a $\frac{1}{2}''$ diam. roller ended follower.

5. Design a radial cam to give a 30 mm rise and fall to the following data. Shaft 15 mm diam., minimum diam. 25 mm. S.H.M. for 100°; dwell 20°; fall 20 mm 100° U.V.; dwell 20°; fall 120° Uniform Acceleration and Retardation; suitable for a 10 mm diam. roller ended follower, offset 10 mm.

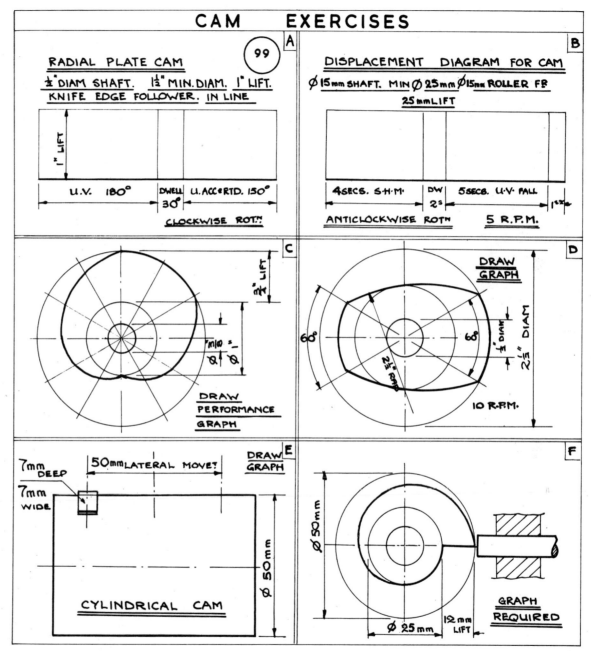

45

99A. Cam Problems

A. A radial plate cam profile is shown. Draw the performance graph of the cam.

B. The shape shown is to be used as a radial plate cam. A roller ended follower, $\frac{5}{8}''$ diam. is used upon the cam. Draw the modified profile of the cam to suit this condition. Draw the performance graph.

C. Draw the cam profile shown, suitable for a $\frac{1}{2}''$ diam. roller follower. Draw the performance graph.

D. The performance graph of a plate cam is shown. Draw the cam profile which would give such an effect.

Describe the performance of the four cases shown above, (a) in terms of degrees, (b) in terms of time per revolution.

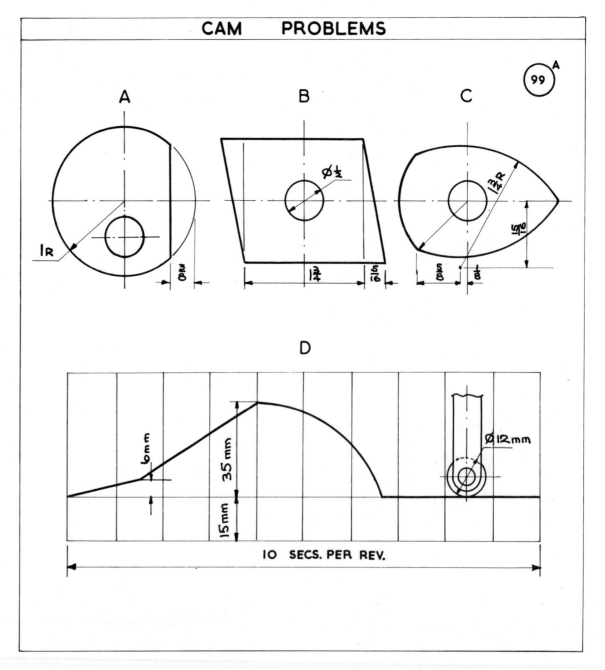

CAM   PROBLEMS

99 A

A

B

C

D

10 SECS. PER REV.

HELICES AND GEARING

Glissette

Roulette

Involute

Spirals

Spur Gearing

Bevels

Worms

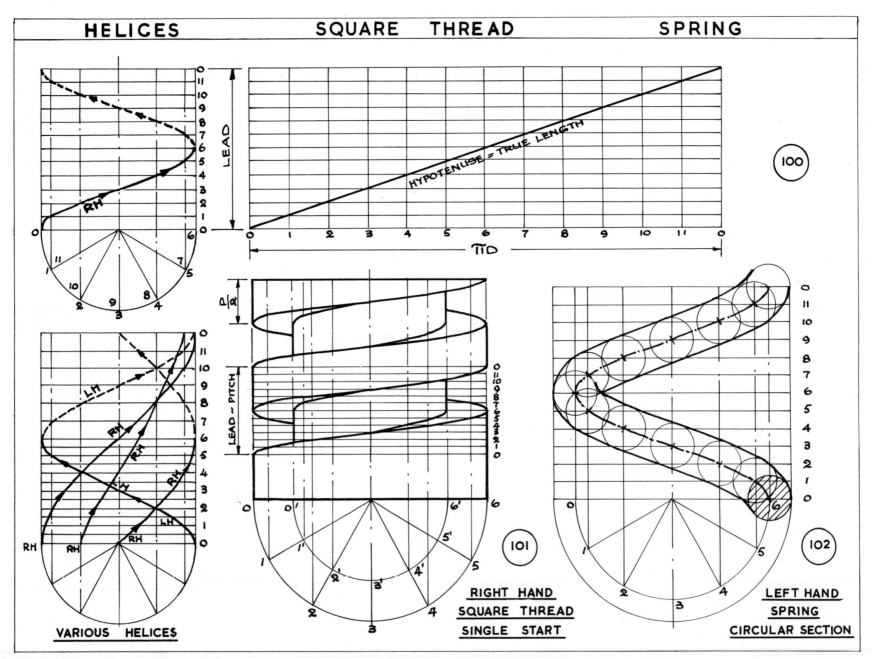

HYPOTENUSE = TRUE LENGTH

LEAD

πD

100

RH

LH

RH

RH

RH

RH

RH

RH

LH

VARIOUS HELICES

LEAD ~ PITCH

D/2

101

**RIGHT HAND
SQUARE THREAD
SINGLE START**

102

**LEFT HAND
SPRING
CIRCULAR SECTION**

100. The Helix The locus of a point which moves around the surface of a cylinder so that its axial and radial speeds are in a constant ratio. The axial advance is the lead, the pitch is the distance from a point on the helix to a similar point for one revolution—centre to centre of the screw thread for example. In a one start thread the lead and pitch are the same, for a two start, the pitch is half the lead. The projection of the helix is a straight line. The helix may be right-hand or left-hand. To draw the helix, project the twelve generators to intersect corresponding twelve divisions of the lead. The curve drawn is a simple harmonic motion curve, see cams. Various helices are shown.

101. Right-Hand Square Thread single start. Pitch equal to lead. Thread section a square P/2 side. Generate two helices, one for the root of the thread, and one for the crest.

102. Left-Hand Spring, circular section. The helix to be plotted is the centre of the circular section. The lines of the spring are then drawn tangential to the construction circles.

103. Two Start R.H. square section thread. Mark off the lead on the cylinder blank, make the pitch half the lead, the thread section is a square P/2 side. Draw the helices, four for the crest, four for the root.

104. Three Start L.H. square thread, pitch is one-third lead. The helices rise to the left-hand, six for the crests, six for the roots. See also Worms in Spur Gearing.

Exercises

1. Draw a single start square section screw thread of 3″ outside diam., lead 1″. Complete three turns, L.H.

2. Draw one and a half turns of a spring, 8 cm mean diameter and 5 cm lead, R.H., made from 15 mm diam. wire.

3. Draw a two start square section thread, L.H. 25 mm pitch, outside diam. 10 cm. Complete two turns.

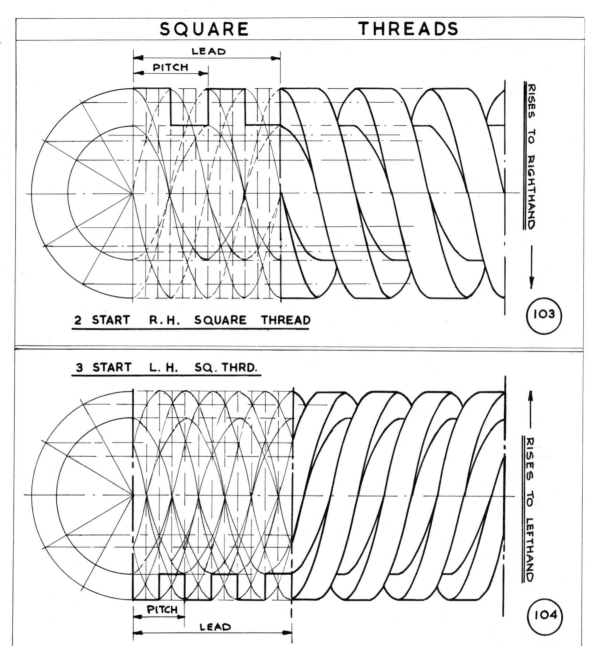

SQUARE THREADS

LEAD

PITCH

RISES TO RIGHTHAND

2 START R.H. SQUARE THREAD

103

3 START L.H. SQ. THRD.

RISES TO LEFTHAND

PITCH

LEAD

104

105. Glissette When a line slides between two other lines which are fixed, the locus of a point on the sliding line is called a glissette. The lines may be at any angle and may be straight or curved. The example shows a straight line moving through six positions between two straight lines which are at right angles. Locus of P is the glissette.

106. Roulette When one curve rolls upon another without sliding, the locus of a point P on the rolling curve describes a roulette. The points can be plotted by tracings or cutouts of the two curves in progressive positions.

107. Involute of a Circle If a straight generating line rolls round a circle the locus of a point on the line is the involute of the circle. The Evolute is the circle itself. The involute curve is used in obtaining the form of spur gear teeth. To construct the involute, divide the generating circle into twelve parts and draw the straight line equal to the twelve divisions and thus the circumference. Draw the first tangent and make it the length of one division. Draw the tangent at 2 and make it two divisions in length. Proceed in this manner until the twelve tangents have been drawn. Draw a fair curve through the points, this is the involute. If a cord is unwound from a cylinder, a pencil placed in the end loop will trace the involute curve.

108. The Involute of a Square Draw the square, draw the quadrants as shown from the corners. The radius increases by the length of a side at each quadrant.

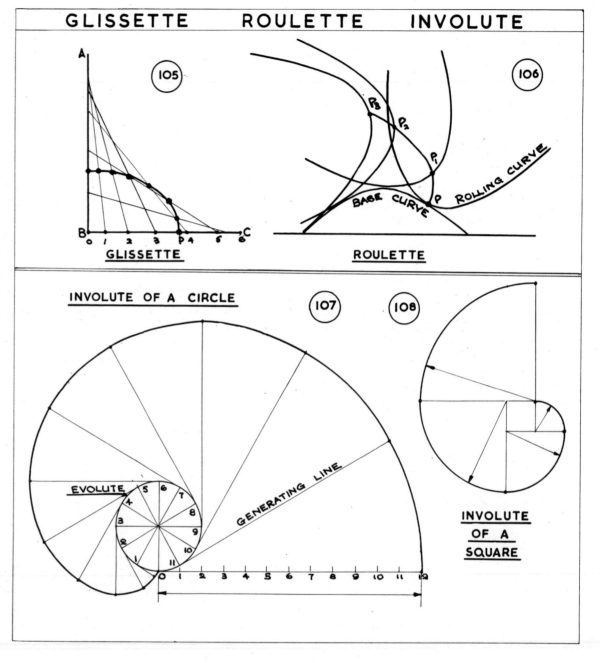

Spirals

The locus of a point moving continuously in one direction along a line which is rotating about one of its ends.

109. Archimedean Spiral Given a pole P and limiting vectors, A length and B length. Draw A and B in line from the pole, divide A into twelve parts, and draw twelve radial vector lines. Draw arcs to cut the radials progressively as in the diagram and construct a fair curve to pass through the points of the spiral.

The normal and tangent to any point on the spiral can be drawn if the value of C (which is a constant) be found by the formula

$$C = \frac{r-a}{\theta \text{ (in radians)}}$$

where r is the radius length of the point
a is the initial radius vector length
θ is the vectorial angle in radians
$(1° = ·017453 \text{ rad.}).$

Draw C at right angles to the line joining P and the point on the spiral. Draw the normal from the end of C to the point; the tangent is at right angles to the normal.

110. Logarithmic Spiral Given the angle between the vectors, the length of the first vector, and the ratio of the vector lengths, 5:4. From the pole C draw the radial lines of the vectors at the angle given. Obtain the lengths of the vectors by drawing x=vector, $y=\frac{5}{4}\times$ vector, angle 45°, and by arcs and parallels obtain the progressive vector lengths to be used on the numbered radials. A fair curve through the points gives the spiral as shown in the diagram.

111. Conical Spiral Divide the vertical height of the cone into the same number of parts as the base circle. Join the plotted points with a fair curve.

Exercises

1. Draw the involute of a 2″ diam. circle.

2. Draw an Archimedean spiral whose vectors are 50 mm and 15 mm.

3. Draw the logarithmic spiral given two vectors $\frac{3}{4}$″ and 1″ at an angle of 40°.

4. Draw the spiral of a cone of 2″ height and 3″ diam. base circle.

5. Draw the spiral of a hemisphere of 9 cm diameter.

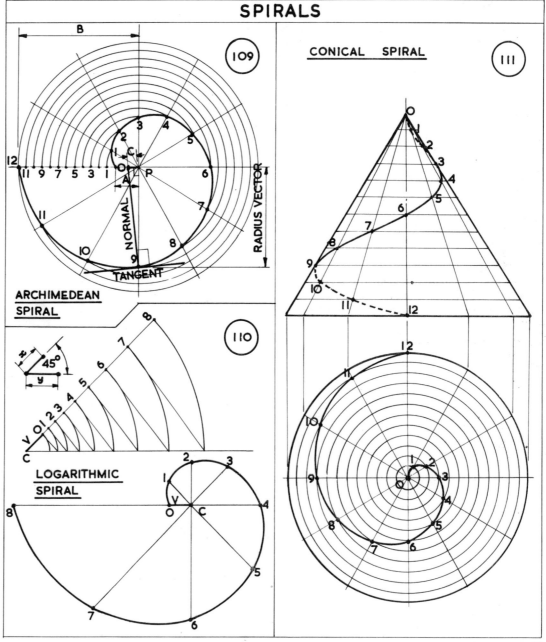

SPIRALS

ARCHIMEDEAN SPIRAL — 109

LOGARITHMIC SPIRAL — 110

CONICAL SPIRAL — 111

112. Involute Spur Gears The involute curve of a circle is used in the profile shape of the involute gear tooth. The base circle is the generating circle and is generated tangentially to the pressure line.

113. Terms used in Spur Gearing.

Pitch circle diameter	P.C.D.
Pitch point	
Line of action	
Pressure angle	$20°$ or $14\frac{1}{2}°$
Addendum	a
Dedendum	d
Clearance	c
Circular pitch	p
Circular tooth thickness	$p/2$
Number of teeth	T
Diametral pitch	D.P.
Module	m
Base circle diameter	B.C.D.

The circular pitch is the distance centre to centre of the teeth measured on the pitch circle.
Diametral pitch is the number of teeth per inch of pitch circle diameter. A ratio.
Module is the reciprocal of the diametral pitch.

It is usual to give the pressure angle; the number of teeth, and the diametral pitch of the gear to be drawn. Other essential proportions and sizes can be found by the use of the formulae shown.

114. First draw the touching pitch circles.
Draw the pressure line through the pitch point.
Draw the base circles tangential to the pressure line.
Draw the involute curves on the base circles.
Step off the pitch points on the pitch circles.
Draw the addendum and dedendum circles.
Draw the involute curves through the pitch points.
Complete the shape of the teeth by drawing the root radii.
Approximate compass curves may be used in drawing the teeth, using one of the methods shown later.

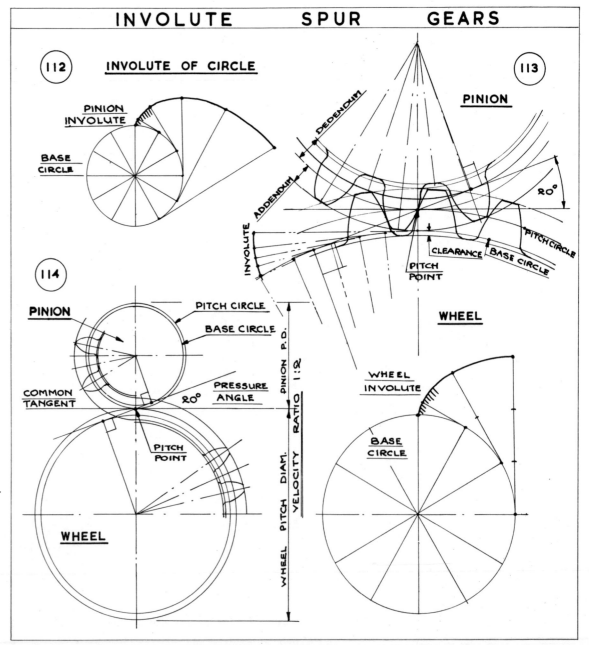

INVOLUTE SPUR GEARS

115. A Worked Example of Spur Gears.

A wheel and pinion ratio 2:1; number of teeth, wheel 30; pinion 15; pressure angle 20°; diametral pitch 3·14; circular standard pitch 1″; module 8·085 mm (·315″).

Apply the given data in the formulae to obtain the pitch diameter, the addendum, clearance and dedendum.

Draw the touching circles, pitch diameters.

Draw the common tangent to give pitch point.

Draw the pressure line, 20° to the common tangent and through the pitch point.

Construct the base circles tangential to the pressure line.

Draw the addendum circle, and the dedendum (root) circle.

Construct the involute curves for the wheel and pinion on the base circle of each.

Step off the half-pitch distances along the pitch circles.

Draw the involute tooth curve through these points to form the teeth. Use a tracing for reproduction, or an approximate compass curve of the involute.

Complete the tooth shape by root curve, radius W/7.

Compass curves are allowed to conventionally represent the true involute curve, the radius being one-eighth of the pitch diameter. The centres for the arcs are on the circle which passes through the intersection of the two arcs D/4 and D/8 shown in the diagram. The fillet radius at the tooth root is one-seventh of the widest space between teeth on the tip circle. An approximate compass curve can also be taken from the involute curve itself by bisecting a chord as shown in the next diagram.

Spur Gear Formulae

$$D.P. = \frac{T}{P.C.D.} \quad also \quad P.C.D. = \frac{T}{D.P.}$$

$$p = \frac{\pi \times P.C.D.}{T}$$

$$Addendum = \frac{1}{D.P.}$$

$$Clearance = \frac{p}{20}$$

$$Module = \frac{1}{D.P.}$$

INVOLUTE GEAR EXAMPLE

EXAMPLE :

GIVEN :

GEAR RATIO 2:1

NO. OF TEETH 30:15.

PRESSURE ANGLE 20°

STANDARD C.P. 1″

D.P. 3·14   MODULE 8·085 ᵐ/ₘ

			WHEEL	PINION
PRESSURE ANGLE		20° OR 14½	20°	20°
NO. OF TEETH	T	D×DP	30	15
DIAMETRAL PITCH	DP	$\frac{T}{D}$	3·14″	3·14″
PITCH CIRCLE DIAM	D	$\frac{T}{DP}$	9·55″	4·78″
ADDENDUM	A	$\frac{1}{DP}$ or ·3183p	·301″	·301″
CIRCULAR PITCH	p	$\frac{\pi}{DP}$	1″	1″
CLEARANCE	c	$\frac{p}{20}$	·05″	·05″
DEDENDUM	d	$\frac{A+c}{or}$ ·3683p	·351″	·351″
TOOTH THICKNESS		$\frac{p}{2}$	·5″	·5″
MODULE	m	$\frac{1}{DP}$	8·085 ᵐ/ₘ (·315″)	8·085 ᵐ/ₘ (·315″)

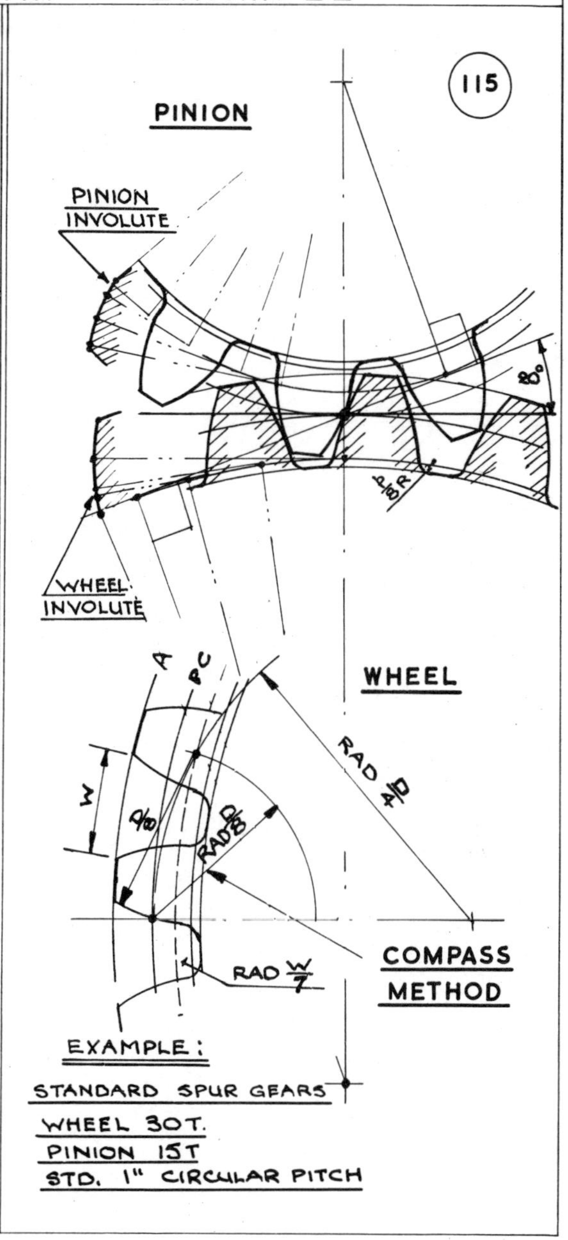

PINION

PINION INVOLUTE

WHEEL INVOLUTE

WHEEL

COMPASS METHOD

RAD $\frac{D}{4}$

RAD $\frac{D}{8}$

RAD $\frac{W}{7}$

EXAMPLE :

STANDARD SPUR GEARS

WHEEL 30 T.

PINION 15 T.

STD. 1″ CIRCULAR PITCH

(115)

116. A Worked Example of Spur Gearing.

From the given data calculate the Pitch circle diameter, the addendum, the dedendum, the tooth thickness and the circular pitch.

Proceed in the same order as the previous example. Two equal gears have been shown, but a velocity ratio of 3:4 could be taken as an exercise.

117. The Form of the Involute Rack is shown.

The involute curve taken to infinity becomes a straight line, the proportions of the rack are as the meshing gear, the slope of the flanks of the rack teeth are the same angle as the pressure angle given.

Exercises

1. Draw the spur gear indicated by the following data; standard 1″ circular pitch, pressure angle 20°, teeth 22.

2. Two spur gears mesh to give a velocity ratio of 1:5, and have a pressure angle of 20°, D.P. 3·5″, and the pinion has 16 teeth. Draw two teeth of the pinion meshing with three teeth of the wheel.

3. Draw a spur gear which has 30 teeth, a pressure angle of 20°, and a pitch circle diameter of 9·55″. Only five teeth need be completed, draw the teeth profile by using a tracing from the involute curve.

4. The centre distance of two gears is 9″, the pinion radius is 2½″. The D.P. is 3·2″, pinion teeth 16. Draw the gears meshing, completing three teeth of each gear.

5. A pinion has 12 teeth, pitch diameter 3·8″, pressure angle of 20°, and meshes with an involute rack. Draw three complete teeth of the pinion meshing with the rack, of which five teeth should be completed.

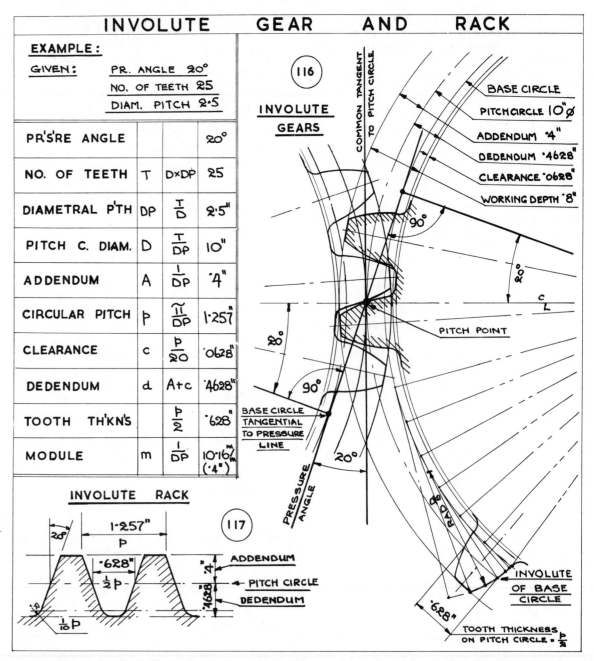

INVOLUTE GEAR AND RACK

EXAMPLE:

GIVEN:

PR. ANGLE		20°
NO. OF TEETH		25
DIAM. PITCH		2·5

PR'SRE ANGLE			20°
NO. OF TEETH	T	D×DP	25
DIAMETRAL P'TH	DP	$\frac{T}{D}$	2·5″
PITCH C. DIAM.	D	$\frac{T}{DP}$	10″
ADDENDUM	A	$\frac{1}{DP}$	·4″
CIRCULAR PITCH	P	$\frac{\pi}{DP}$	1·257″
CLEARANCE	c	$\frac{P}{20}$	·0628″
DEDENDUM	d	A+c	·4628″
TOOTH TH'KN'S		$\frac{P}{2}$	·628″
MODULE	m	$\frac{1}{DP}$	10·16^{mm} (·4″)

INVOLUTE GEARS

BASE CIRCLE
PITCH CIRCLE 10″ø
ADDENDUM ·4″
DEDENDUM ·4628″
CLEARANCE ·0628″
WORKING DEPTH ·8″

PITCH POINT

BASE CIRCLE TANGENTIAL TO PRESSURE LINE

COMMON TANGENT TO PITCH CIRCLE

PRESSURE ANGLE

INVOLUTE OF BASE CIRCLE

TOOTH THICKNESS ON PITCH CIRCLE = $\frac{P}{2}$

INVOLUTE RACK

1·257″ P

·628″ ½P

⅒P

ADDENDUM
PITCH CIRCLE
DEDENDUM

·4″
·4628″

54

118. Mitre Bevel Gears The diagram shows the conventional method of representing two equal bevel gears whose shafts are arranged at 45°, and for this reason are called mitre gears.

The axis lines of the shafts are drawn first, followed by the pitch and back cone lines. The teeth section can now be inserted.

119. Reduction Bevel Gears The ratio of the reduction between pinion and wheel is in the same relationship as the pitch circle diameters.

120. Worm and Wormwheel A reduction gearing between two shafts whose axes are at right angles. The toothed wheel is driven by the cylindrical worm whose screw surface is an involute helicoid of single or multistart thread form. The worm tooth section is that of the involute rack (see 117), and the tooth section of the wormwheel is a suitable meshing involute spur gear, shown in detail in No. 122. The reduction ratio is as the number of teeth in the wheel as to the starts in the worm, i.e. a 30-tooth wheel driven by a single start worm gives a reduction of 30:1, whereas a 30-tooth wheel driven by a three-start worm gives a 10:1 ratio. The worm is usually made from case hardened steel or normalised high carbon steel, and the wheel is made from phosphor bronze or cast iron.

Proportions

d = Pitch diam. of worm
D = Pitch diam. of wheel
p_n = Normal pitch of worm
p_a = Axial pitch of worm
a = Addendum of worm. $0\cdot5\,p_n$
b = Dedendum of worm
t = Number of threads in worm
T = Number of teeth in wormwheel
λ = Lead angle of worm (tan $\lambda = tp_a/\pi d$)

Useful guides: make $d=0\cdot2D$; also $a+b$ not greater than $d/2$. Further reference to B.S. No. 721.

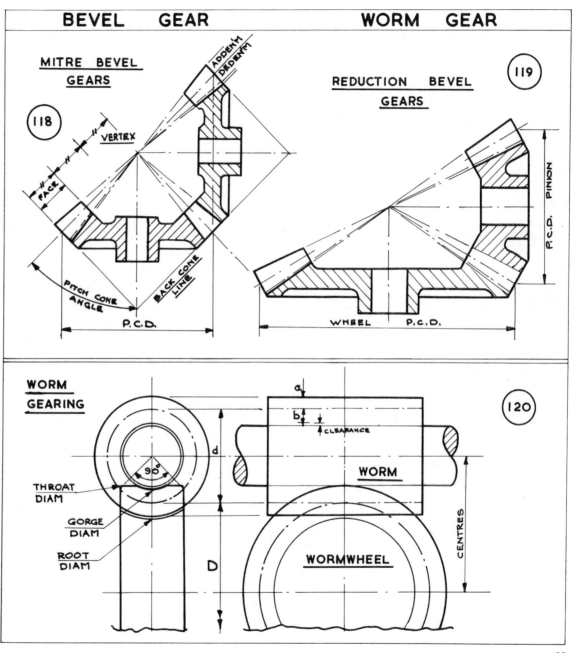

BEVEL   GEAR   WORM   GEAR

MITRE BEVEL GEARS

118

VERTEX

FACE

PITCH CONE ANGLE

BACK CONE LINE

P.C.D.

ADDEN'M
DEDEN'M

REDUCTION   BEVEL GEARS

119

P.C.D. PINION

P.C.D. PINION

WHEEL   P.C.D.

WORM GEARING

120

THROAT DIAM

GORGE DIAM

ROOT DIAM

90°

d

D

a

b

CLEARANCE

WORM

WORMWHEEL

CENTRES

121. Bevel Gears The diagram shows details of a pair of mitre bevel gears, 15 teeth, diametral pitch 3·14, pitch circle diam. 3·4″.

Draw the pitch cones and the back cone lines. Draw the generating tooth form, projecting the pitch circle from the pitch angle line. The spur tooth form is drawn conforming to the proportions given on previous pages. The addendum and dedendum lines can now be swung back to the back cone line and joined to the vertex point. The shape of the teeth can be projected as in the given partial elevation by projection from the section and by lines radial from the elevation centre to the generating tooth form as shown.

Exercises

1. Draw the pinion of a pair of bevel gears which give a reduction of 2:1; pinion has 10 teeth, standard circular pitch 1″, P.C.D. 3·183″. Complete two teeth.

2. Draw the sectional view of a pair of bevel gears pitch angle 60°, reduction 1:1·25. Pinion has 12 teeth, pitch diam. 3·82″. Circular pitch 1″.

3. A pair of bevel gears work at right angles, the pinion has 15 teeth, gear ratio 5:6, D.P. 3·14″. Draw the sectional views of the gears, show the generating tooth form, and project three teeth of the pinion.

4. Draw one tooth of a spur gear, 20 teeth, P.C.D. 6·366″, 1″ circular pitch. Use this form to give the shape of the teeth of a mitre bevel gear.

5. Draw the pinion of a pair of bevel gears the axes of which are at 85°, P.C.D. 3·82″, teeth 12. Complete the two full elevations 2 × F.S.

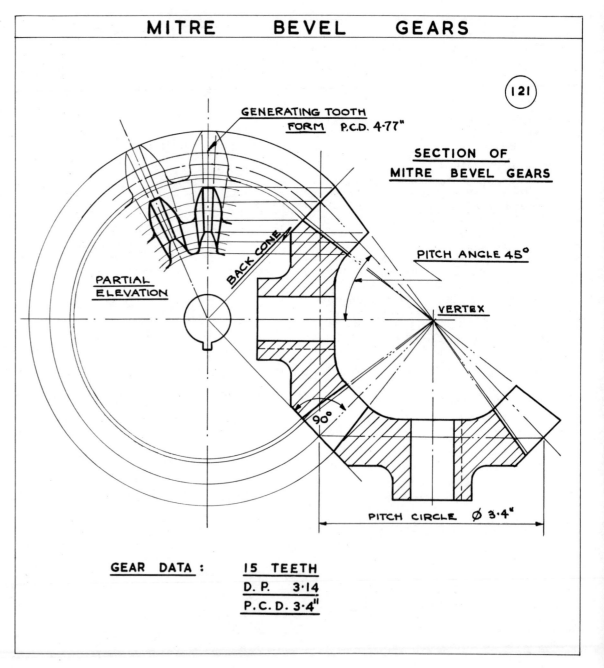

MITRE BEVEL GEARS

121

GENERATING TOOTH FORM P.C.D. 4·77″

SECTION OF MITRE BEVEL GEARS

PITCH ANGLE 45°

BACK CONE

VERTEX

PARTIAL ELEVATION

90°

PITCH CIRCLE Ø 3·4″

GEAR DATA :   15 TEETH
D. P.   3·14
P.C.D. 3·4″

122. Worm and Wormwheel The diagram shows projections of a 1″ lead, single start, worm meshing with a 30 tooth standard 1″ circular pitch involute gear wormwheel. P.C.D. 9·55″ diam.

Draw the mid-section of the wormwheel first by the methods previously shown—pitch circle, tangent, pressure line, base circle, tooth spacing, tooth shape by approximate compass form.

Draw the mid-section of the worm teeth to involute rack shape, the tangent line forming the base line. The rack form has been shown on a previous page. Make the centre core of the worm not less than twice the depth of the teeth, this enables the cylinder of the worm to be found. Draw the construction for the single start helices, 1″ lead. The worm is shown completed in the diagrams.

123. Train of Gears A simple train of gears is shown, the ratio of reduction or increase is equal to the number of teeth in the final gears, the intermediate gears being idlers which only affect the direction of rotation.

In the compound train, the reduction or increase is obtained by multiplying the teeth of the drivers together and dividing into the number obtained by multiplying the driven teeth together. Rotation may also have to be considered, usually effected by the introduction of an equal idler at the beginning.

Two involute gears and rack are shown in oblique projection.

Further specialised information on gears may be obtained from B.S. 436, B.S. 545, B.S. 721, B.S. 2519.

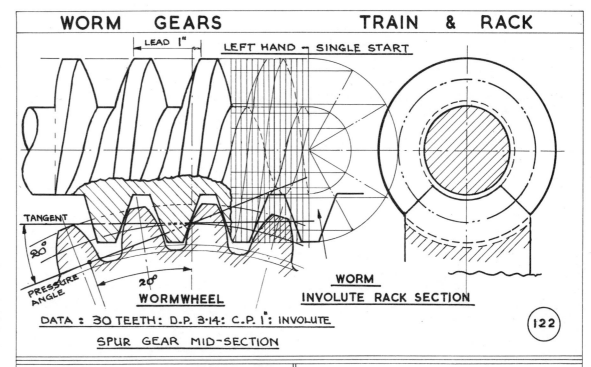

WORM GEARS   TRAIN & RACK

LEAD 1″   LEFT HAND – SINGLE START

TANGENT

20°

PRESSURE ANGLE   20°

WORMWHEEL   WORM
INVOLUTE RACK SECTION

DATA : 30 TEETH: D.P. 3·14: C.P. 1″: INVOLUTE

SPUR GEAR MID-SECTION

(122)

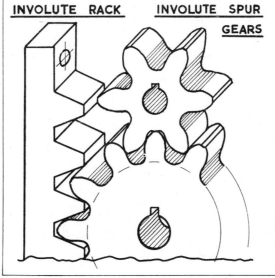

INVOLUTE RACK   INVOLUTE SPUR
GEARS

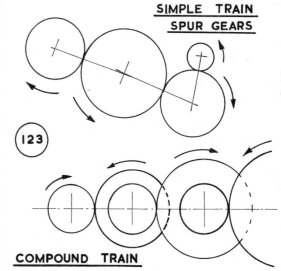

SIMPLE TRAIN
SPUR GEARS

(123)

COMPOUND TRAIN

123A Gear Angles and Forms: Worm. The diagram shows a pitch helix given a worm lead angle of 30°. Draw the centre line (axis) of the worm, draw the semicircle D = Pitch circle diameter. Divide the semicircle into six parts, step off chordal distances 0′, 1′, 2′, 3′ along the base line as shown. Perpendiculars cut the slope line giving the heights to be projected to the generators from 0, 1, 2, 3. The helix can be drawn through the intersection points. Note the spiral angle, also the lead length.

Spur teeth simple involute form, can be shown schematically by projection lines which are parallel to the shaft axis of the gear.

Helical gears single helix, can be indicated by slope lines at the same angle as the helix angle.

End thrust is engendered when single helical gears work under load, and this is usually absorbed by ball or roller races.

Double helical gears are indicated in the third diagram. Gears having this form cancel out endthrust.

Note that when drawing meshing gears of the last two cases, the slopes indicating the teeth will lie to the opposite hand in engagement.

Exercises

1. Draw the pitch helix of a worm, lead angle of 45°. The diameter of the pitch circle is $2\frac{1}{2}''$.

2. Draw helix diagrams where the lead angle is (*a*) 60°, (*b*) $22\frac{1}{2}°$; these angles are frequently used. Use the same pitch circle diameter, $2\frac{1}{2}''$.

3. Draw the three gear forms shown in schematic form, showing in each case a pinion of 2″ diam. meshing with a $3\frac{1}{2}''$ diam. gearwheel.

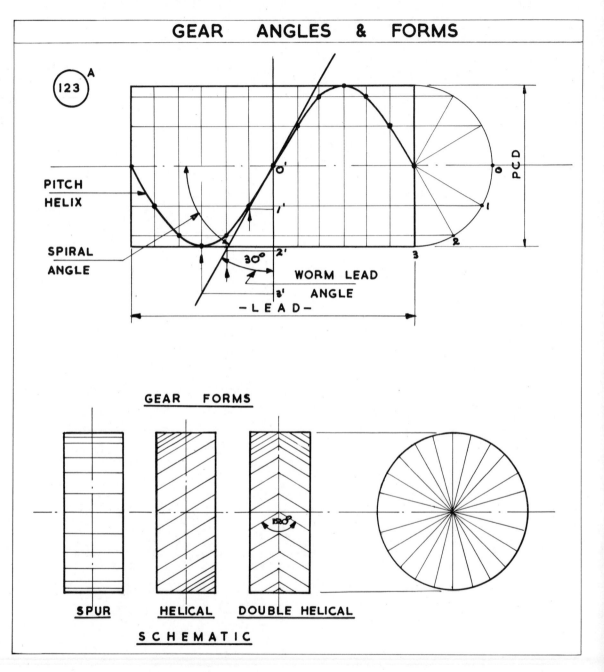

58

FORCES AND FRAMEWORKS

Triangle of Forces

Vector and Stress Diagrams

Beam Moments

Funicular and Link Polygons

Frameworks

124. Resolution of Forces

A. If two concurrent co-planar forces act upon a point, they must be equal in magnitude (amount) and direction (line of action) if equilibrium (balance) is to be maintained.

B. An unstable couple is set up if the forces are equal but not opposite. Rotation ensues.

C. Two parallel forces may be opposed by a single force equal in magnitude and parallel in direction.

D. Two and two equal parallel forces in a state of equilibrium.

E. Simple example of a common beam centrally loaded by ten units balanced by left and right reactions R_L and R_R of 5 units each. Weight of beam ignored.

F. Offset loading of simple beam by 12 units. The reactions R_L and R_R are found by dividing the load in the ratio 2:6 shown in the diagram as the division of the length of the beam between the supports. Notice that the greater reaction is at R_L in this case, owing to the load being nearer that end. The weight of the beam, so many weight units per foot, would be divided equally between the two reactions.

125. G. If two concurrent co-planar forces act at an angle on a point P, their *resultant* (a single force equal to the two forces) may be found graphically by drawing a scale triangle (or a parallelogram) of forces. Represent the two forces by vector lines, length to scale, showing magnitude, direction (angle) as given, to form two sides of the triangle. The third side, closing the triangle, represents the resultant force in magnitude and direction (angle).

The equilibrant—a force equal to the resultant in magnitude and direction (line of action)—acts on the point in the opposite direction (pull or push) to the resultant. In the vector diagram, the arrows must 'flow round' the diagram; note how vectors 1 and 2 must be related to obtain the 'flow'.

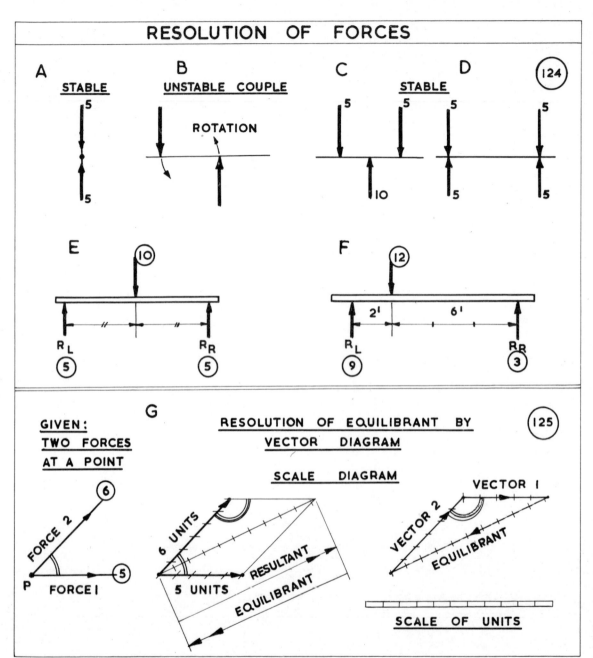

126. Several Concurrent Forces at a point. A. Four forces are shown acting at a point P. A fifth unknown force, the equilibrant, is needed to maintain balance. This is found by drawing a vector diagram showing the known forces to a suitable scale in both magnitude and direction to the given data. The fifth, closing side of the polygon, shows the equilibrant as a vector quantity, and its magnitude and direction to scale. If the forces are arranged so that the arrows 'flow' round the polygon, the push or pull action of the equilibrant is indicated.

127. B. A further example of forces acting at a point. When the vector polygon is being drawn, the closing line showing the equilibrant, crosses the vector F_2, but F_1, F_2 and F_3 must be drawn as shown in order that the arrows can 'flow' round the polygon.

128. C. Non-Concurrent Co-Planar Forces Given the direction and magnitude of four forces which are not acting on the same point, find the force necessary to establish equilibrium.

Label the spaces between the forces as shown using capital letters—this is known as 'Bow's Notation'. Construct a vector polygon as above of the known forces to scale and close the polygon so obtaining the magnitude and direction of the equilibrant F_5.

Letter the points of the vector diagram by small letters a, b, c, d, e. Thus the vector ab represents the force F_1 lying between A B of four units magnitude and direction. Join $abcde$ to any pole O in the vector diagram.

In space B, in the space diagram, draw a line parallel to Ob; in space C, a line parallel to Oc, completing the diagram with lines Od, Oe, Oa. Intersection of Oe and Oa will give a point through which F_5, the equilibrant, may be drawn parallel to ae.

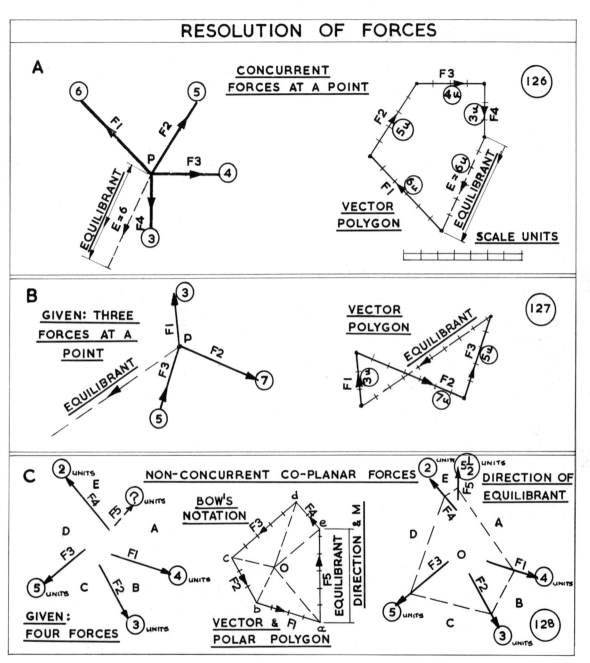

RESOLUTION OF FORCES

61

129. Beam Moments The resolution of forces acting upon a beam may be found graphically using Bow's Notation in a space diagram; a load line and a polar polygon; a funicular or link polygon; all to suitable scales.

A beam is shown with loads of 100, 200 and 300 units, arranged at intervals shown. Find the reactions at R_L and R_R.

Graphic Solution Draw the beam as shown to a scale of $\frac{1}{4}'' = 1$ ft. Label the spaces A, B, C, D, to conform with Bow's Notation. Draw the load line scale $\frac{3}{4}'' = 100$ units, showing 100 units a to b; 200 units b to c; 300 units c to d. Fix any pole O. Join $abcd$ to pole O, forming the polar polygon. Begin the link polygon by drawing aO parallel to aO in the polar polygon. Draw bO, cO, dO as shown to give y in the link polygon. Join xy with eO. Draw eO in the polar polygon parallel to eO in the link polygon, cutting the load line to give the magnitude of the reactions R_L and R_R.

Beam Reactions by Calculation A 'moment' is obtained by multiplying the load by its distance from the point being considered.

Moments at R_L are three, 100×6; 200×14; 300×20, total 9,400. These may be ton/ft; lb/ft; kilo/metre, etc). Moments at R_R are, $R_R \times 24$, that is, 9,400. Therefore R_R is 391·66 weight (load) units; and R_L will be 208·34 weight units.

Notice that the equilibrant, or single force required to maintain the beam in equilibrium, is shown acting at a point vertically above z in the link polygon, balancing the total load of 600 units (beam weight has been ignored in this example).

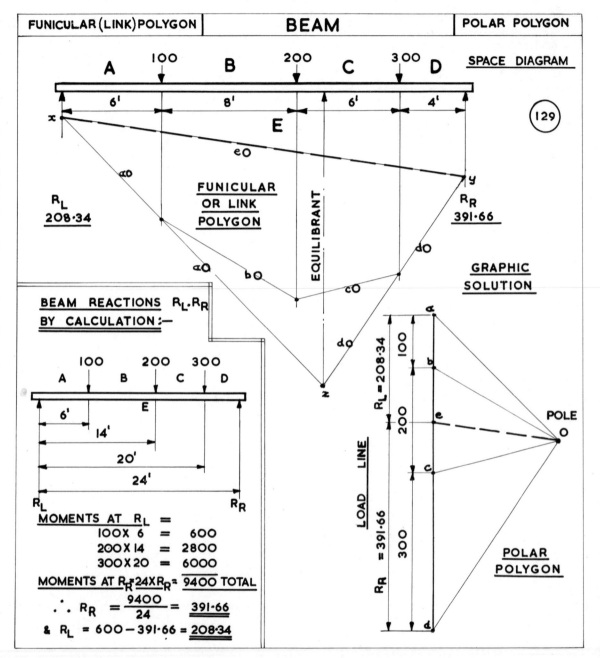

130. Shear Force If a loaded beam were cut at any point between the supports, the beam would collapse as the forces acting upon it would be unbalanced.

The Shear Force is obtained by calculating the upward forces (as moments), and the downward forces (as moments) at the section, and subtracting. Positive Shear, (+ve), is indicated when the force tends to turn the beam in a clockwise direction; Negative Shear, (−ve), when the beam tends to turn in an anti-clockwise direction.

The Bending Moment The force required to maintain a cut beam in a state of equilibrium is known as the Bending Moment, and is found by calculating the force at the point of section.

Graphical Solution Draw the beam and loads, to a convenient scale. Label as in Bow's Notation. Draw the load line to scale, fix pole O and draw the polar polygon. Draw the shear force diagram. Construct the link or funicular diagram. Find the reactions of the beam supports, R_L and R_R.

Scale of the shear force diagram is the scale used for the forces in the load line, of the polar diagram.

Scale of the bending moment diagram is given by:
Bending Moments in units feet

= space diagram scale in feet
× polar distance in inches
× load line scale in units (wt).

Moments of Beams The reactions R_L and R_R of beams in the examples are resolved (*a*) by the graphic method, using a load line; a polar polygon; a link or funicular polygon; (*b*) by calculation, weight × distance =moment.

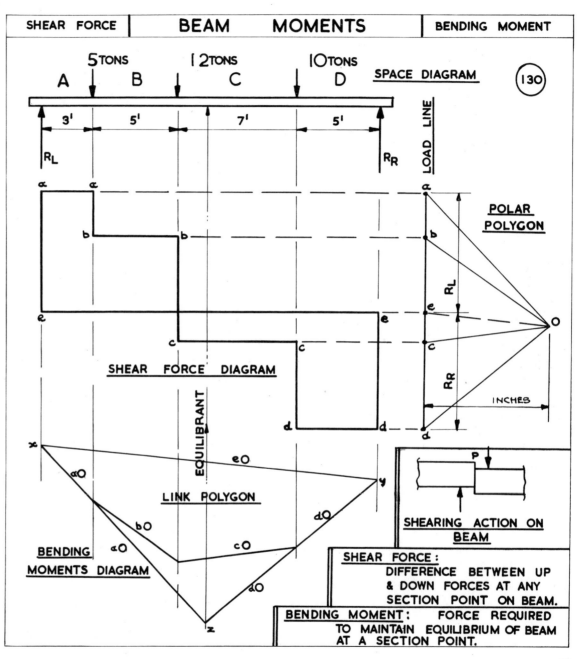

SPACE DIAGRAM

POLAR POLYGON

SHEAR FORCE DIAGRAM

EQUILIBRANT

LINK POLYGON

BENDING MOMENTS DIAGRAM

INCHES

SHEARING ACTION ON BEAM

SHEAR FORCE: DIFFERENCE BETWEEN UP & DOWN FORCES AT ANY SECTION POINT ON BEAM.

BENDING MOMENT: FORCE REQUIRED TO MAINTAIN EQUILIBRIUM OF BEAM AT A SECTION POINT.

131. Simple Central Load The topload on the beam of 400 units requires 200 units at each of the reactions R_L and R_R to maintain equilibrium.

132. Offset Load A load of 600 units rests on the beam 1′ from R_L and 5′ from R_R. The diagram shows how the moments are obtained.

133. Multiple Loading A beam has loads of 200, 300 and 100 units weight at the intervals shown. By the graphic method, notate the spaces A, B, C, D, and draw the load line to scale, *abcd*. Join *abcd* to any pole O. From *x* draw the funicular or link polygon. Join *x* and *y* in *e*O. Draw *e*O in the polar polygon parallel to *e*O, and read off the magnitudes of R_R and R_L to scale, on the load line. The diagram shows the calculation of the moments also.

134. Overhung Beam The diagram shows details of the loading of an overhung beam. Notate the spaces A, B, C, D, E, and draw the load line to scale *abcde*. Join *abcde* to a pole O. Draw the link polygon beginning at *x*, obtain *f*O. Draw *f*O in the polar polygon. Read off the magnitude of R_L and R_R to scale on the load line. Calculation of the moments by arithmetic is also shown.

135. Cantilever Beam Draw the beam shown to a suitable scale. Notate the spaces A, B, C, D, E. Draw the load line *abcd* to scale. Join *abcd* to a pole O. Draw the link polygon beginning at *x*. Notice how line *c*O is taken to the vertical line CD, *d*O returning to give *y* on the reaction line R_R. Line *e*O must lie between the reaction lines. Draw *e*O in the polar polygon parallel to *e*O. Read off the magnitude of R_L and R_R to scale on the load line.

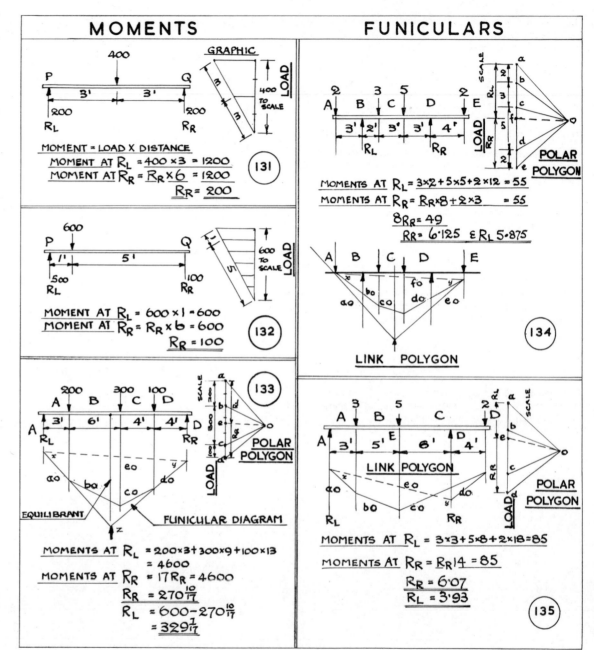

MOMENTS

MOMENT = LOAD × DISTANCE

MOMENT AT R_L = 400 × 3 = 1200

MOMENT AT R_R = R_R × 6 = 1200

R_R = 200

(131)

MOMENT AT R_L = 600 × 1 = 600

MOMENT AT R_R = R_R × 6 = 600

R_R = 100

(132)

(133)

MOMENTS AT R_L = 200×3 + 300×9 + 100×13
= 4600

MOMENTS AT R_R = 17R_R = 4600

R_R = $270\frac{10}{17}$

R_L = $600 - 270\frac{10}{17}$
= $329\frac{7}{17}$

FUNICULARS

MOMENTS AT R_L = 3×2 + 5×5 + 2×12 = 55

MOMENTS AT R_R = R_R×8 + 2×3 = 55

8R_R = 49

R_R = 6·125 & R_L 5·875

LINK POLYGON

(134)

MOMENTS AT R_L = 3×3 + 5×8 + 2×18 = 85

MOMENTS AT R_R = R_R14 = 85

R_R = 6·07

R_L = 3·93

(135)

136. Simple Framework Draw the framework to a suitable scale. Letter the spaces to comply with Bow's Notation. Draw the force line *abc* to show 4 units of load to a suitable scale. Construct the link or funicular diagram by drawing *ad* parallel to A D and *bd* parallel to B D. Draw *dc* parallel to C D. Magnitude of force in members A D and B D can be read off to scale. The direction of stress is indicated in the three smaller diagrams and show that the two members are in compression and are *struts*. The member C D is in tension and is a *tie*; its magnitude can be scaled off from the link diagram also. Notice how the arrows follow round in one direction in the vector diagrams, to indicate direction of stress.

137. Framework Uniformly loaded beam. Reactions are equal and half the total load. Draw the force line to a suitable scale, *abcd*. Draw the link or funicular, start at *af*, *dj*, parallel to A F, D J. Other vectors are parallel to corresponding members. Notice how *fe* cuts the force line to give the reactions to scale of 500 units. Scale off the magnitudes of members. Draw in arrows to represent direction of stress in members—strut or tie. Struts can also be shown by heavy lines, and ties by thin lines, or arrowed.

138. Framework A more complicated example with unequal loading. Notate the spaces. Draw the force line *abcd* to scale. Join *abcd* to pole O. Draw the beam funicular to find the magnitudes of R_L and R_R; starting from *x* draw *a*O parallel to *a*O of the polar polygon, *b*O parallel to *b*O, and so progress to *y*. Join *xy* with *e*O. Draw *e*O parallel to *e*O, this gives the magnitudes of R_L and R_R to scale units. Draw the vector diagram, *af* parallel to A F, etc. The completed force diagram enables the stress of members to be read on the scale. The detailed vector diagrams show the direction of stress in the members, and is the key to later problems.

Note: 'I' is omitted in Bow's Notation, avoiding confusion with numeral one.

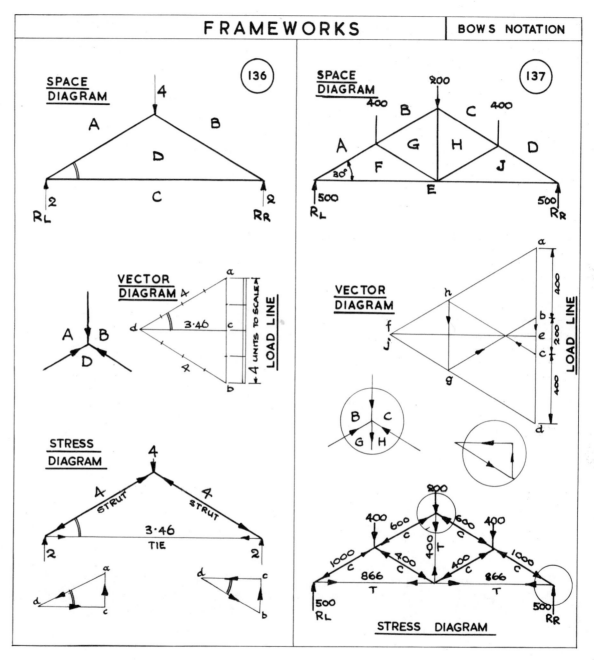

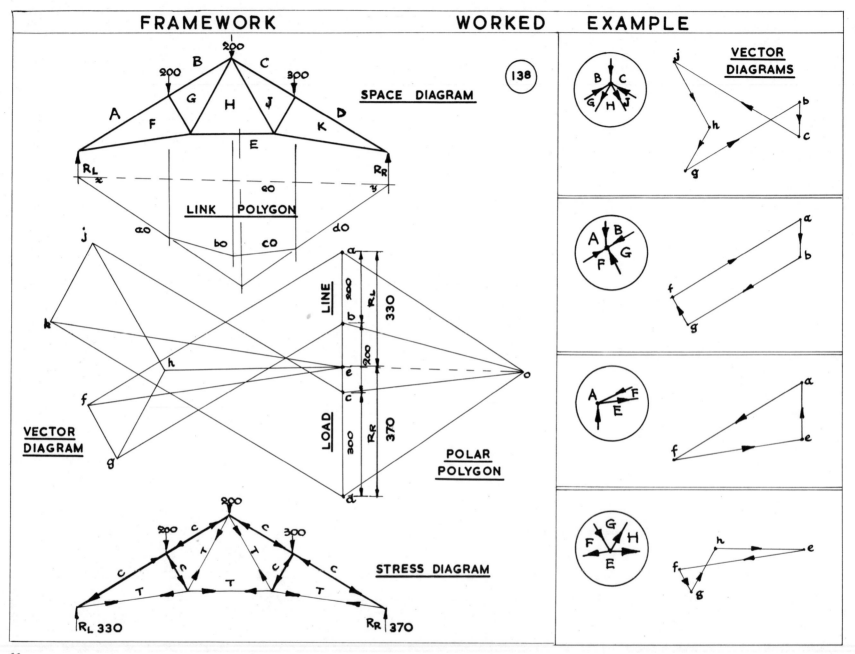

SPACE DIAGRAM

LINK POLYGON

VECTOR DIAGRAM

LINE

LOAD

POLAR POLYGON

STRESS DIAGRAM

VECTOR DIAGRAMS

139. Warren Framework Draw the frame to a suitable scale, letter the spaces to Bow's Notation. Draw the load line *abc* to a suitable scale and parallel to the direction of the two load forces. Join *abc* to pole O. Construct the link polygon, join *x* and *y* to give *d*O. Draw *d*O in the polar diagram giving *d* and the reactions R$_L$ and R$_R$.

Draw the vector diagram by vector lines parallel to the members in the space diagram, *ae* from *a* and parallel to A E; *de* from *d* and parallel to D E; *bf* from *b* and parallel to B F; *dg* from *d* and parallel to D G; *cg* from *c* and parallel to C G. The length of the vector lines in the vector diagram are measured to the same scale used in the load line and give the magnitude of the stress. The direction of the stress is given by the arrows and indicates whether the member is a tie or strut.

140. Warren Framework, second example. Draw the frame to a suitable scale. Letter the spaces. Draw the load line, to scale, *abc*. Draw the link polygon, obtain *xy*. Draw *d*O parallel to *d*O (*xy*) and obtain the reactions R$_L$ and R$_R$. Construct the vector diagram, by drawing parallel vectors to the respective members of the framework, *aj*, *dj*, as a starting point. Measure the vectors to the load scale line to obtain their magnitude, show the direction by arrows. Indicate whether tie or strut.

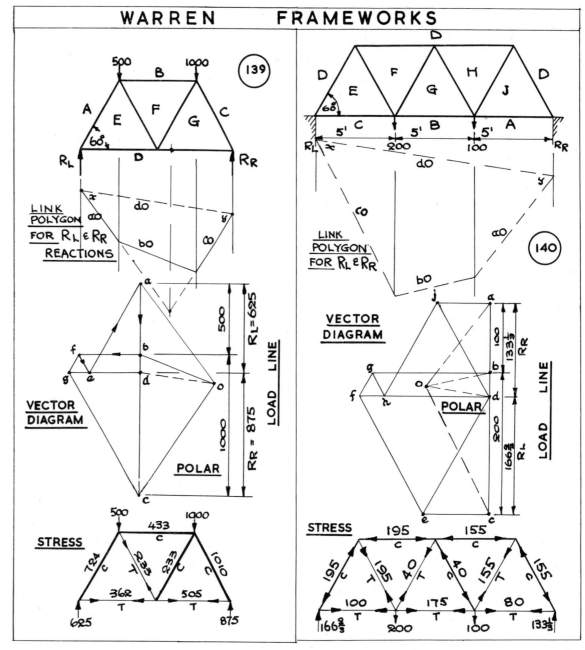

WARREN FRAMEWORKS

141. Box Type Framework Draw the given frame. Notate the spaces as in Bow's Notation, A, B, C, D, E, F, G, H, J. Draw the load line to a suitable scale showing the loads *ab* 50 units; *bc* 100 units; *cd* 100 units. Join *abcd* to the pole O, forming the polar polygon. Draw the link polygon beginning at *x*, making *b*O parallel to *b*O; *c*O parallel to *c*O. Join *xy* giving *e*O. Draw *e*O in the polar polygon parallel to *e*O, giving the magnitude of R_R and R_L, the reactions.

Draw the vector diagram, begin with *ch, jh, hg, gf,* parallel to their respective members in the space diagram.

Notice that B F and E J have no vector length and zero stress.

The resultant and equilibrant can be found, if required, by drawing *a*O from *x* parallel to *a*O; and *d*O from *y* parallel to *d*O from the polar polygon. The line of the equilibrant is drawn through *z* and is parallel to the load forces.

The magnitude and type of stress in members of the framework may either be shown in a stress diagram, or in a table showing name of member, load, and type of stress. A specimen table is shown.

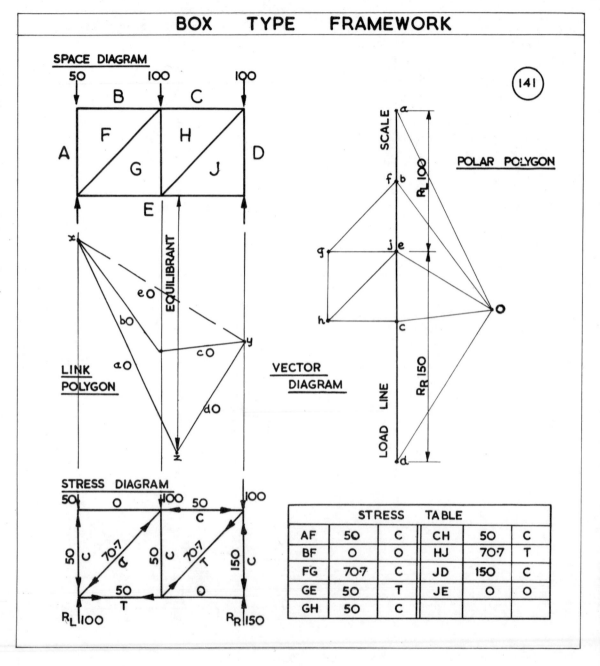

BOX TYPE FRAMEWORK

STRESS	TABLE				
AF	50	C	CH	50	C
BF	O	O	HJ	70·7	T
FG	70·7	C	JD	150	C
GE	50	T	JE	O	O
GH	50	C			

142. Wind Loads on a Framework On the previous examples, loads with a vertical line of action have been shown. Roof trusses have to stand wind pressures, and this example shows angled loading which is often found.

Draw the frame and letter the spaces. Draw the load line to a suitable scale showing the three forces and parallel to the action line of the forces. Draw the polar polygon joining *abcd* to pole O. Begin the link polygon at *x*, and draw *a*O, *b*O, *c*O and *d*O. Notice *y* is on a vertical from the point of R_R. Draw *e*O joining *xy*, and transfer to the polar polygon to find the reactions R_R and R_L. The line of the reaction at R_L will be vertical and P is at the intersection of this line and BC force load line. Join P to R_R to give the line of action of R_R.

Draw the vector diagram to obtain the loads and stresses in members of the frame. Show these forces in the stress diagram.

143. Resolution of Two unknown Forces acting at a point. Draw the two known vectors *ad* and *dc*, draw *ab* and *cb* parallel to the forces P and Q. Join *abcd* to a pole O, to complete the polar polygon.

144. Non-parallel Forces acting on a Beam. Draw the given beam and forces. Letter the spaces. Draw the load line to show magnitude and direction of the forces to scale. Join *ad* giving the equilibrant. From *x* draw the link polygon; the equilibrant is drawn parallel to *ad* and through *z*.

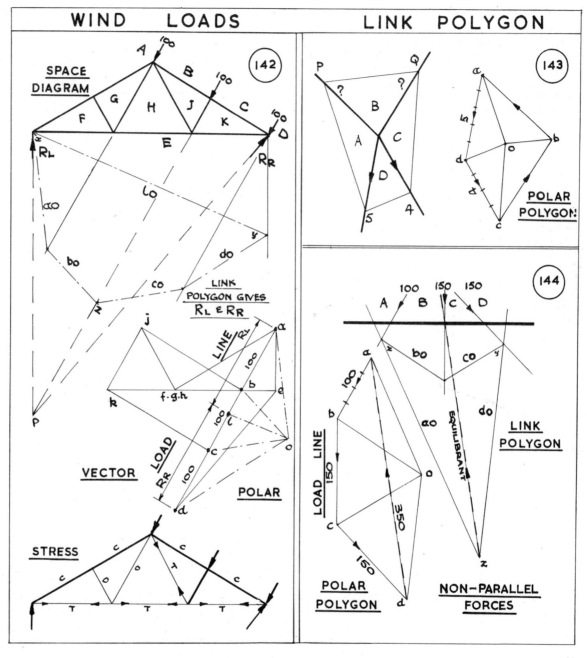

WIND LOADS   LINK POLYGON

SPACE DIAGRAM

142

VECTOR

POLAR

LINK POLYGON GIVES R_L & R_R

STRESS

143

POLAR POLYGON

144

LINK POLYGON

POLAR POLYGON

NON-PARALLEL FORCES

LOAD LINE

145. Canted Truss A canted frame with suspended loads is shown. Draw the frame and notate the spaces. Draw the load line *bcde* to scale and direction. Join *bcde* to pole O forming the polar polygon. Draw the link polygon. Draw *a*O, and find *a*O in the polar polygon. Construct the vector diagram, read off the magnitude of stress in the vectors using the same scale as the load line. Draw the stress diagram as an exercise.

146. Framework with top, wind and suspended loads. Draw the frame and notate the spaces. Draw the load line *bcde′f′*, join to pole O. Draw the link polygon. Transfer *a*O parallel to polar polygon giving R_L and R_R to scale on load line. Draw the vector diagram. Draw the stress diagram. Tabulate the stresses in the members.

147. Roof Truss with wind and dead loads. Draw the frame and notate. Draw the load line *bcdef* and join points to pole O, to construct the polar polygon. (Note that the wind and dead load at BC, CD, DE, must be resolved into one force each by a local triangle or parallelogram of forces; this gives the line of direction of load line for the force.)

Draw the link or funicular polygon, note that the line of R_L is vertical. Draw R_L line vertically downward from *b* to meet *a*O. Draw *af* to give R_R. Measure R_L and R_R on the scale of the load line. Construct the stress diagram and indicate the direction and magnitude of force in each member. Tabulate the result.

148. Angled Roof Truss with wind and dead loads. Draw the frame to suitable scale, and notate the space diagram. Draw the load line to scale, join *bcdefg* to pole O. Draw the link polygon. R_L line of action is vertical, from *b* to *a*, *ag* gives R_R. Draw the vector diagram, measure the vectors on the load line scale to obtain the magnitude in the member. Draw the stress diagram. Tabulate the results.

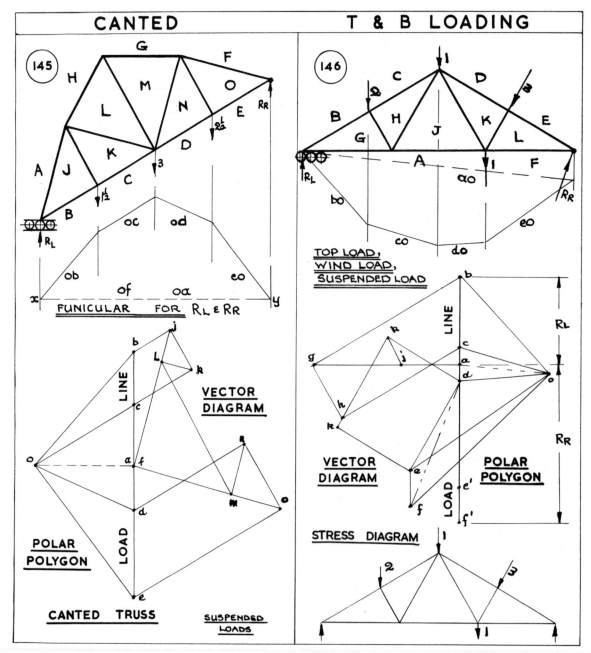

CANTED — T & B LOADING

FUNICULAR FOR R_L & R_R

POLAR POLYGON

VECTOR DIAGRAM

CANTED TRUSS

SUSPENDED LOADS

TOP LOAD, WIND LOAD, SUSPENDED LOAD

VECTOR DIAGRAM

POLAR POLYGON

STRESS DIAGRAM

70

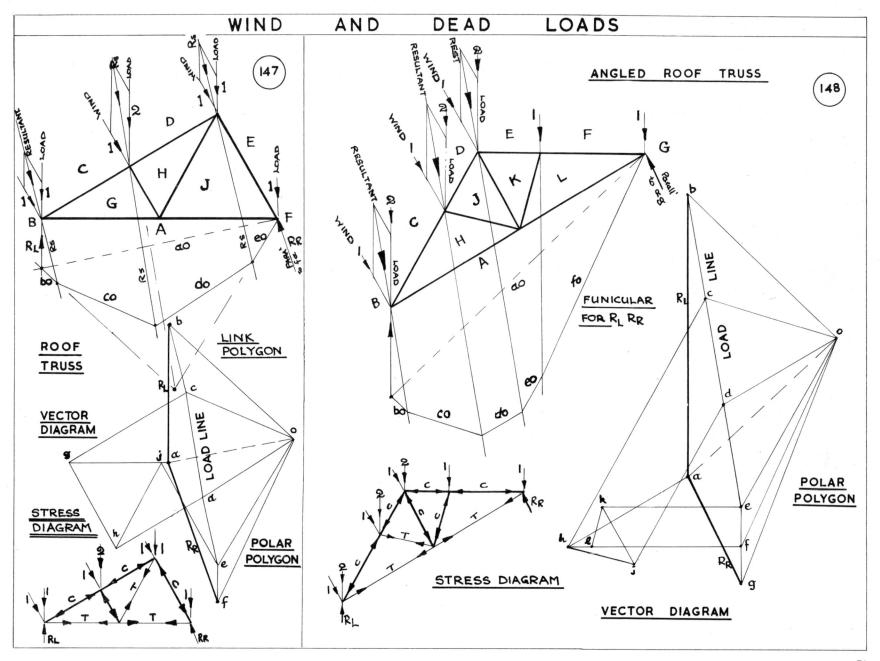

ROOF
TRUSS

VECTOR
DIAGRAM

STRESS
DIAGRAM

LINK
POLYGON

POLAR
POLYGON

ANGLED ROOF TRUSS

FUNICULAR
FOR R_L R_R

STRESS DIAGRAM

VECTOR DIAGRAM

POLAR
POLYGON

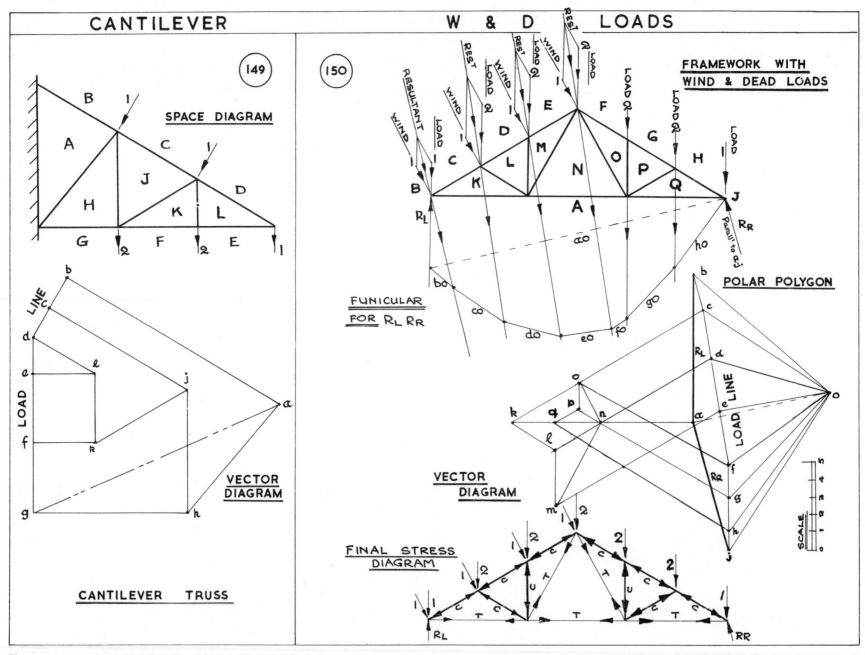

CANTILEVER

149

SPACE DIAGRAM

LINE

LOAD

VECTOR DIAGRAM

CANTILEVER TRUSS

150

FRAMEWORK WITH
WIND & DEAD LOADS

FUNICULAR
FOR R_L R_R

POLAR POLYGON

LOAD LINE

VECTOR
DIAGRAM

FINAL STRESS
DIAGRAM

SCALE

149. Cantilever Truss A wall framework having top angled loads and a vertical bottom load is given. Draw the frame and notate the spaces. Draw the load line *bcdefg* to a suitable scale and parallel to the forces shown. Draw the vector diagram, begin with *dl* and *el*; then *fk* and *lk*; *cj* and *kj*; *gh* and *jh*; *ba* and *ha*. Join *ag* to find the magnitude and direction of the reaction of A G. Draw the stress diagram as an exercise.

150. Framework with Wind and Dead Loads Draw and notate the frame. Find a single resultant force for each of the wind and dead loads at B C, C D, D E and E F, by drawing a small scale parallelogram of forces. Extend the resultant lines obtained downward to enable the link polygon to be drawn later.

Draw the load line *bcdefghj* to scale magnitude and direction. Join these points to pole O to make the polar polygon. Draw the funicular polygon to obtain *aO*, and the reactions R_L and R_R. R_L is vertical. Draw the vector diagram and obtain the magnitude and stresses in the frame members.

151. Jib Crane An overhung frame is given. Draw the frame and notate. Begin the vector diagram with the load line *ab* vertical and showing 500 scale units.

Draw vectors *ac*, *bc*; then *ae*, *ce*; *ed*, *bd*; *bf*, *df*; *fg*, *bg*. Notice *bj* is zero.

Draw the final stress diagram, show the magnitude and stress in the members.

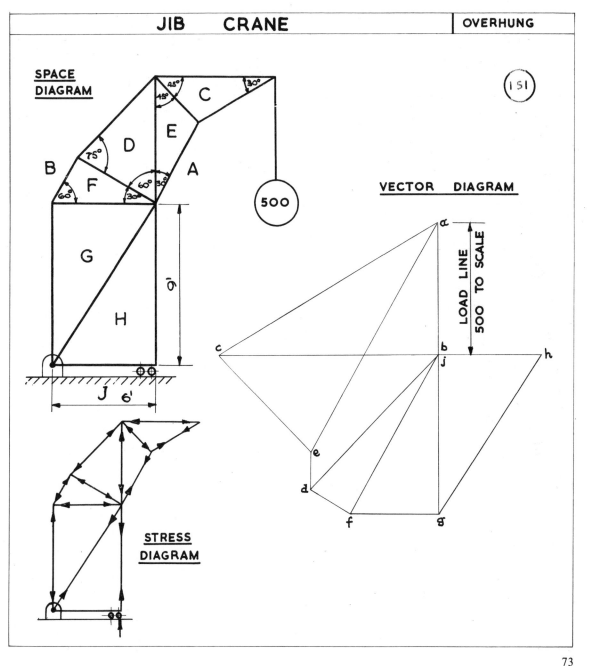

SPACE DIAGRAM

VECTOR DIAGRAM

STRESS DIAGRAM

152. Two Forces Unknown in a Structure In the wall bracket frame shown, the two unknown forces are P and Q. Find their magnitude and direction.

Draw the given frame, notate the spaces. Begin the vector diagram by drawing the load line *ab* to 100 scale units. Draw vector *ad* from *a* parallel to AD. Draw vector *bd* from *b* and parallel to BD, to intersect in *d*. All force lines must pass through the same point, shown at *x*, and this must lie on the vertical line AB in the space diagram and at the intersection of a line coincident with BC. Join *x*Q to give the direction of force Q. Complete the vector diagram by drawing *bc* from *b* and parallel to BC; *cd* from *d* and parallel to CD. The magnitude of the stress in the members in the frame is found by measuring the respective vector on the same scale as the load line *ab*, and the compression or tension by arrow 'flow'.

153. Hinged Frame The frame is hinged at R_R and rests on rollers at R_L to allow for movement under strain. The direction of the reaction at R_L is assumed to be at right angles to the centre lines of the rollers, in this case vertical. Draw the frame and notate the spaces. Begin the vector diagram with the load line *abc* to scale showing 60 units at 60° to the horizontal, and 30 units on the vertical loadline, as shown. Draw vectors *ad* and *cd* parallel to the load lines, these will give the magnitude of R_L and R_R. Draw vectors *ae* and *be* parallel to AE and BE to complete the diagram. Draw the stress diagram from the information shown in the vector diagram.

Angled forces and load lines Notice that the load lines are always drawn following the load force lines on the frame.

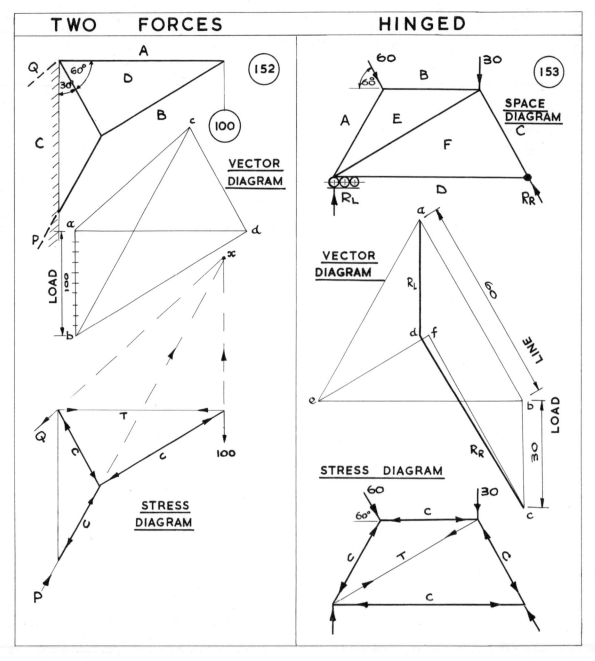

154. Beam Problems Four examples of loaded beams are given. Draw the diagrams to conveniently large scale, notate, and solve graphically, by load line, polar polygon, link polygon, and vector diagram (when required).

155. Six Examples of Frameworks are given, the diagrams should be scaled up to a convenient working size, and solved by drawing load lines, polar polygons, link polygons, vector diagrams and stress diagrams showing the magnitude and type of stress in the members of the frameworks.

The stress in members should also be set out in a table listing the notation of the member, the magnitude of the stress, and whether tie (tension) or strut (compression).

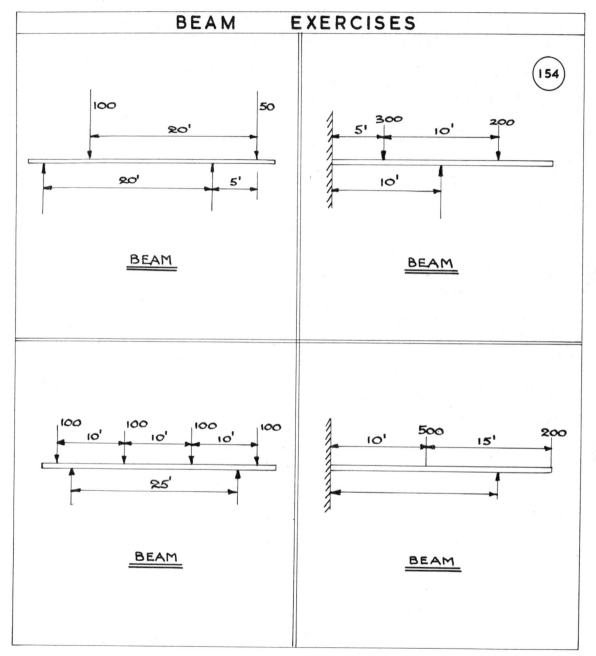

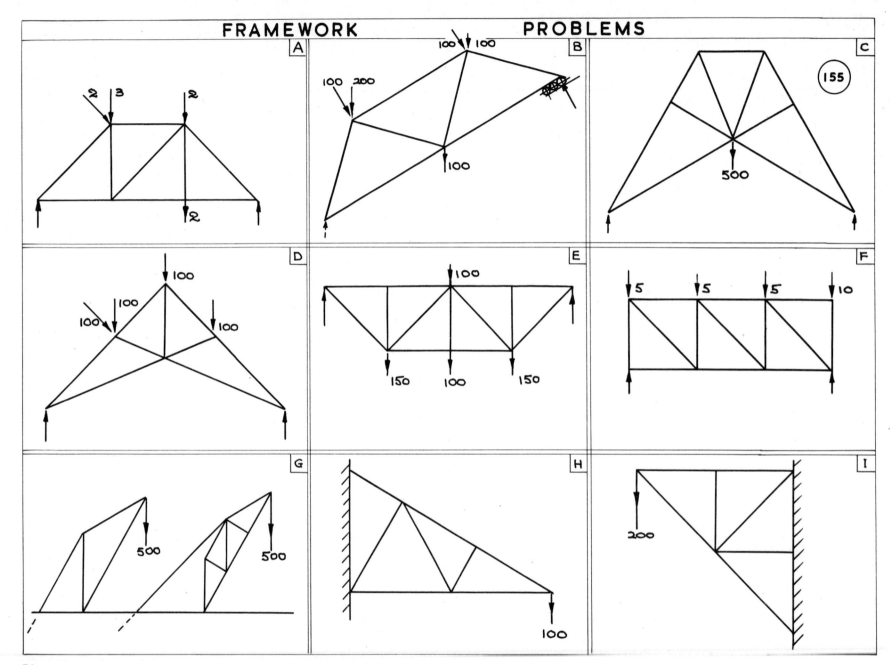

CENTROIDS

Simple

Calculation and Graphic Methods

Integration of Areas

First and Second Moments

Formulae

156. Centroids The centroid of an area is the centre or mean position of all the elements of which the area consists.

Square. Circle. Rectangle. Parallelogram Centroid is the geometric centre.

Triangles The intersection of the medians is the centroid.

Trapezium The method is shown in the diagram.

Quadrilateral *First method.* Divide into two triangles by drawing one diagonal. Draw the medians in the two triangles formed to find C_1 and C_1; join these. Draw the second diagonal, and by medians find C_2 and C_2; join these. The intersection gives C, the centroid.

157. *Second method.* Divide the sides of the quadrilateral into three equal parts. Join the appropriate points to form the second quadrilateral. The diagonals of this quadrilateral intersect in the centroid point.

158. The centroid of a re-entrant quadrilateral can be found by the same method of joining third points to form a rectangle the diagonals of which intersect in the centroid.

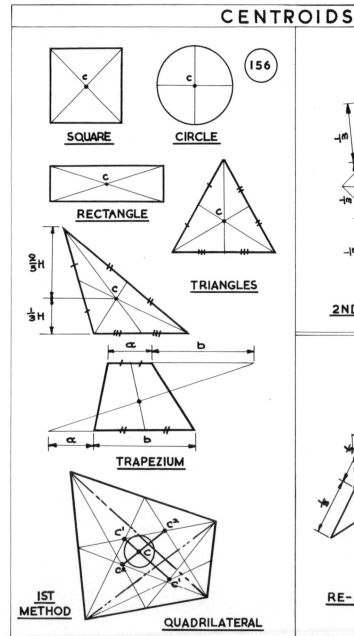

CENTROIDS

SQUARE CIRCLE 156

RECTANGLE

TRIANGLES

TRAPEZIUM

1ST METHOD

QUADRILATERAL

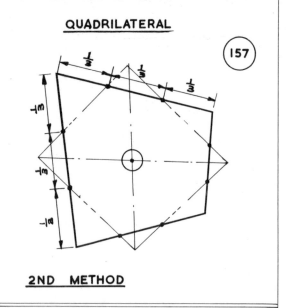

QUADRILATERAL 157

2ND METHOD

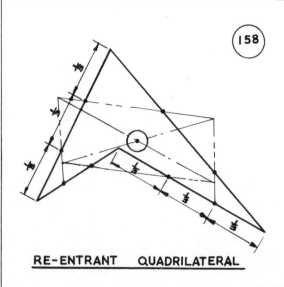

158

RE-ENTRANT QUADRILATERAL

159. **Centroid** of L shaped figure. Divide the figure into two rectangles. Centroid of each is given by diagonals. Join M and L. The centroid lies on this line which should be divided in the ratio of the areas of A and B. Area of A = 15, Area of B = 20. Divide ML in this ratio. Note that the shorter distance is nearer the larger area.

160. **Centroid** of a figure with three rectangular units. Divide the figure into three rectangular units, A, B and C. Proceed as above to find N. Join N to O. Divide N O in the ratio of the areas of A B to C, i.e. 7:4 to give P which is the centroid of all the elements in the figure.

161. Centroids Found by Funicular or Link Polygons Divide the figure again into the three composing rectangles and find the three centroids by drawing diagonals.

A force diagram represented by the line *abcd* drawn to any convenient scale, shows the area value in units of the three rectangles. Join *abcd* to any pole O to obtain the funicular polygon. Draw the three parallel lines from the centroids. Note these are parallel to the force line *abcd*. From *x* draw B*o* parallel to *bo* in the funicular. From its intersection with B C, draw C*o* parallel to *co* in the funicular, to give *y*. From *x* draw A*o* parallel to *ao* in the funicular. From *y* draw D*o* parallel to *do*. The intersection gives *z*, which is the resultant of the three forces. Draw the line of the resultant parallel to the force lines from the centroids.

A second position for a second force diagram and funicular must now be made, the force line *ghij* showing the value of areas BCA, but labelled for clarity G H, H I and I J, in the diagram. Proceed as before to obtain the second resultant line, and it should be clear that the centroid lies on the intersection of the two resultant lines.

Further use of funicular polygons and force diagrams will be found in Frameworks and Beams Diagrams in another section of this book.

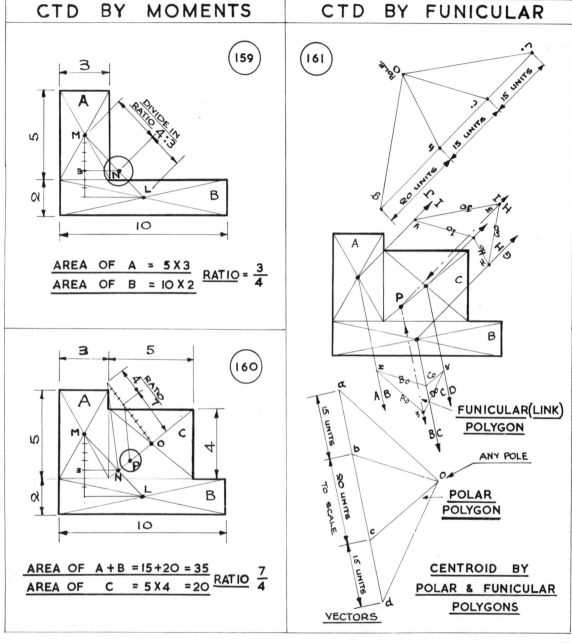

CTD BY MOMENTS

159

AREA OF A = 5 X 3 / AREA OF B = 10 X 2 RATIO = 3/4

160

AREA OF A + B = 15 + 20 = 35 / AREA OF C = 5 X 4 = 20 RATIO 7/4

CTD BY FUNICULAR

161

FUNICULAR (LINK) POLYGON

ANY POLE

POLAR POLYGON

CENTROID BY POLAR & FUNICULAR POLYGONS

VECTORS

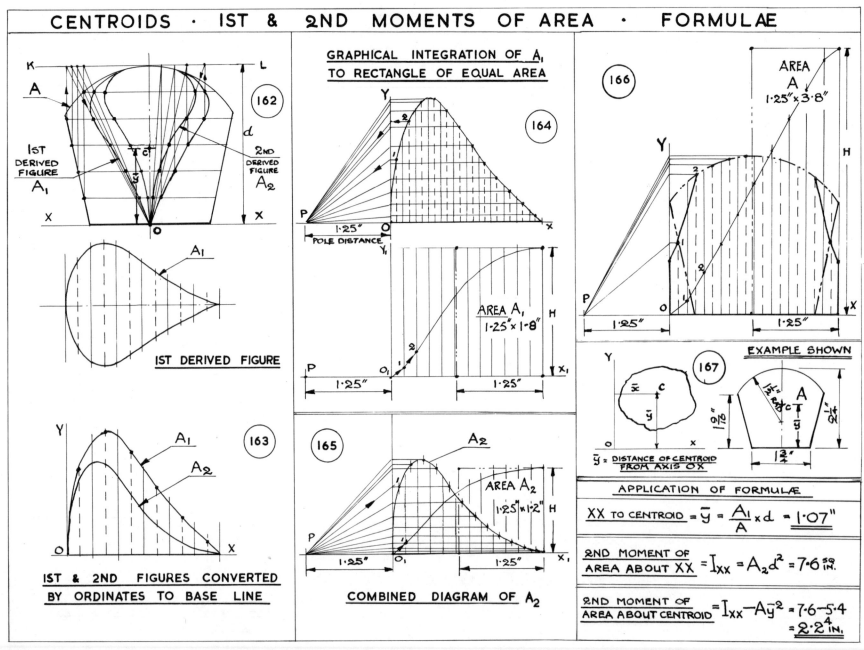

GRAPHICAL INTEGRATION OF A₁
TO RECTANGLE OF EQUAL AREA

162

A
IST DERIVED FIGURE A₁
2ND DERIVED FIGURE A₂

A₁

IST DERIVED FIGURE

163

A₁
A₂

IST & 2ND FIGURES CONVERTED
BY ORDINATES TO BASE LINE

164

POLE DISTANCE

AREA A₁
1·25" x 1·8"

165

A₂

COMBINED DIAGRAM OF A₂

166

AREA A 1·25" x 3·8"

167

ȳ = DISTANCE OF CENTROID FROM AXIS OX

EXAMPLE SHOWN

APPLICATION OF FORMULÆ

XX TO CENTROID $= \bar{y} = \dfrac{A_1}{A} \times d = \underline{1 \cdot 07''}$

2ND MOMENT OF AREA ABOUT XX $= I_{XX} = A_2 d^2 = 7 \cdot 6 \, \tfrac{SQ}{IN.}$

2ND MOMENT OF AREA ABOUT CENTROID $= I_{XX} - A\bar{y}^2 = 7 \cdot 6 - 5 \cdot 4$
$= \underline{2 \cdot 2 \, \tfrac{4}{IN.}}$

162. Centroid of an Irregular Figure The centroid of the shape shown in the example may be found by using the first derived area (A_1) and the second derived area (A_2) called the first and second moments, in the formulae shown in the diagrams. This gives $\bar{y}$ which is the height of the centroid from the axis XX.

Draw the shape standing on the axis XX and draw KL parallel to XX at d height. Divide d height into convenient parts. Where each of these horizontal divisions cuts the shape draw a vertical line up to KL and join to a pole O on XX. Where this line cuts each horizontal a second point is obtained and a fair curve through these gives the first derived area or first moment. Repeat the process to find the second area using the points of intersection of the first area.

The areas of the shape, the first derived area and the second derived area can be found (*a*) by squares, (*b*) by use of ordinates and Simpson's Rule, (*c*) by graphical integration, below:

163. The Shapes of A_1 and A_2 are shown first brought down to an axis OX for convenience. The shape is now divided into equal sections and ordinates drawn in the centre of each section. *Each ordinate height* is now brought in turn to axis OY and joined to a convenient pole P on the extension of OX. *The first two* are numbered on the diagram.

164. Diagram OPY is a funicular and the sloping lines may now be used in the next diagram. Project the section lines down to axis O_1 and X_1. From O_1 draw a sloping line parallel to P_1 in the funicular. The line 1–2 to section two is drawn parallel to P_2 in the funicular. Continue until the line is completed. The area of the shape can now be obtained by multiplying H by PO. By graphical integration the shape has been converted to a rectangle of equal size and area.

165. The Area of the Second derived figure A_2 is shown in a combined diagram.

166. The Area of A the original figure is also shown in a combined diagram.

167. Formulae The values of A, A_1 and A_2 are used in the formulae shown, to find the value of $\bar{y}$, i.e. the height of the centroid above the axis XX.

168. Centroids First and second moments of area.
A = total area of figure.
A_1 = first derived figure.
A_2 = second derived figure.
δA = area of element distance x from OY and y distance from OX.
$\bar{x}$ = distance of centroid from OY.
$\bar{y}$ = distance of centroid from OX.
Moment of area oA about OY = $x\delta A$.
Moment of area oA about OX = $y\delta A$.
Moment of area (total) about OY at centroid = Ax.
Moment of area (total) about OX at centroid = Ay.
Sum of moments about OY = $\sum x\delta A$.
Sum of moments about OX = $\sum y\delta A$.

$$\bar{x} = \frac{\sum x\delta A}{A} \quad \text{and} \quad \bar{y} = \frac{\sum y\delta A}{A}.$$

First moment of area about OY = $x\delta A$.
First moment of area about OX = $y\delta A$.
Second moment of area about OY = $\sum x^2\delta A$, shown Iyy.
Second moment of area about OX = $\sum y^2\delta A$, shown Ixx.
Second moment of area about axis = A (Radius of Gyration)^2.

Distance of centroid from

$$XX = \bar{y} = \frac{A_1 d}{A}.$$

Second moment of area = Ixx = $A_2 d^2$.
Second moment of area about axis through centroid = Ixx − A$\bar{y}^2$.

Graphical Method Summary
1. Draw the figure. By integration obtain its area = A.
2. Draw the first derived figure. Integrate its area = A_1.
3. Draw the second derived figure. Integrate its area = A_2.
4. Calculate distance of centroid from XX, i.e.

$$\bar{y} = \frac{A_1 d}{A}.$$

5. Calculate second moment of area, $A_2 d^2$ = Ixx.
6. Calculate second moment of area about axis through centroid,

$$Ixx - A\bar{y}^2 \quad \text{i.e.} \quad A_2 d^2 - A\bar{y}^2.$$

(Continued on page 83.)

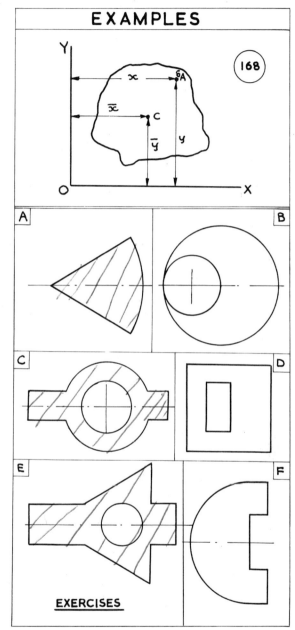

EXAMPLES

168

A   B

C   D

E   F

EXERCISES

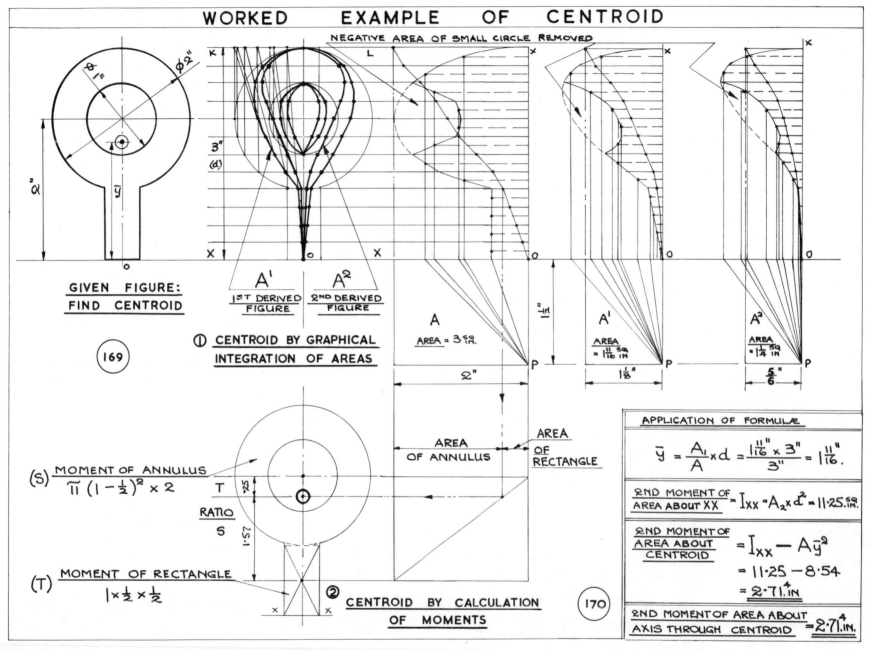

NEGATIVE AREA OF SMALL CIRCLE REMOVED

Ø 2"

Ø 1"

3"
(d)

$\bar{y}$

GIVEN FIGURE:
FIND CENTROID

169

A¹   A²

1ST DERIVED   2ND DERIVED
FIGURE   FIGURE

① CENTROID BY GRAPHICAL
INTEGRATION OF AREAS

A
AREA = 3 SQ. IN.

A¹
AREA = 1 11/16 SQ IN

A²
AREA = 1 1/4 SQ IN

2"   1 1/8"   5/6"

1 11/16"

(S) MOMENT OF ANNULUS / π(1 - 1/2)² × 2

T

RATIO S

.25

1.57

(T) MOMENT OF RECTANGLE / 1 × 1/2 × 1/2

② CENTROID BY CALCULATION OF MOMENTS

170

AREA OF ANNULUS   AREA OF RECTANGLE

APPLICATION OF FORMULÆ

$$\bar{y} = \frac{A_1}{A} \times d = \frac{1\frac{11}{16}" \times 3"}{3"} = 1\frac{11}{16}".$$

2ND MOMENT OF AREA ABOUT XX $= I_{xx} = A_2 \times d^2 = 11.25$ SQ.IN.

2ND MOMENT OF AREA ABOUT CENTROID
$$= I_{xx} - A\bar{y}^2$$
$$= 11.25 - 8.54$$
$$= 2.71 \text{ IN}^4$$

2ND MOMENT OF AREA ABOUT AXIS THROUGH CENTROID $= 2.71$ IN⁴

The centroid could also be found by the intersection of $\bar{y}$ and $\bar{x}$. This would mean all the construction to find distance $\bar{y}$ from X X, followed by a similar construction to find $\bar{x}$ from Y Y.

169. Centroid Worked Example The diagrams show a pierced example fully worked out to obtain the height of $\bar{y}$ (bar y).

Draw the shape standing on X X, and place the pole O on the intersection of X X and the centre line; since the figure is symmetrical the centroid will lie on this centre line.

Draw the first derived figure by the method shown on the previous page, notice that a derived figure must also be drawn for the inner circle.

Draw the second derived figure, this also includes a second figure for the inner circle.

Project the widths of the original figure A; the widths of the first figure, A^1; the widths of the second derived figure A^2 to the area diagrams. Bring the widths to a base line so that ordinates can be easily drawn. Notice that the inner circle area is a negative area, and is deducted from the area being drawn, as shown. Full projections of the areas A, A^1, and A^2 are shown. By graphical integration, condense the areas to rectangles of equal area, a common pole P distance (in this case $1\frac{1}{2}''$), helps in the comparison.

Apply the formula to obtain distance $\bar{y}$ as shown.

170. The position of the centroid is also shown by moments (*a*) by calculation (area × distance from X X), (*b*) by projection of ratio of areas from the original area as shown.

Two Centroid Examples

171. A further pierced example is shown, the first and second derived areas are shown, and the area diagram drawn. The graphical integration of the areas should be drawn as in the previous examples as an exercise.

172. A girder type figure is shown, and the area diagram is also shown. The graphical integration diagrams should now be drawn as an exercise.

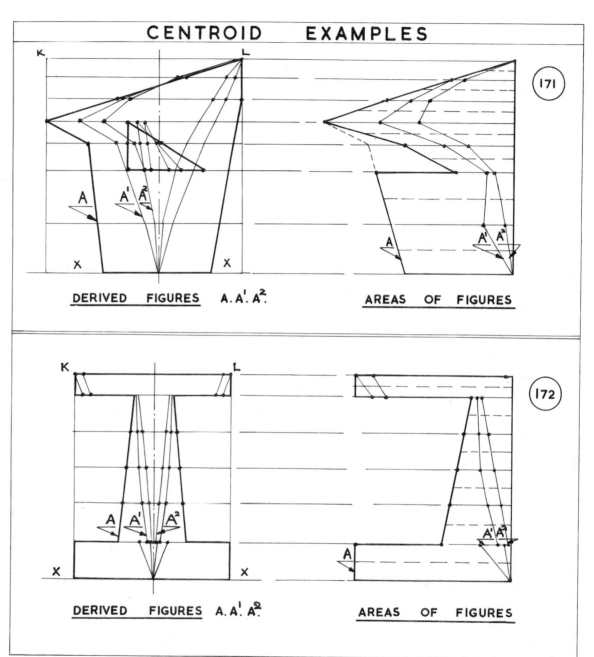

CENTROID EXAMPLES

DERIVED FIGURES A. A'. A².

AREAS OF FIGURES

DERIVED FIGURES A. A'. A².

AREAS OF FIGURES

172A. Centroid Problems Nine shapes are shown suitable for graphical integration, as described in the previous pages.

Draw the shapes larger, scaling up from the printed size; an indication is given in each. Some shapes will be better turned to a more convenient position before integrating to find first and second areas.

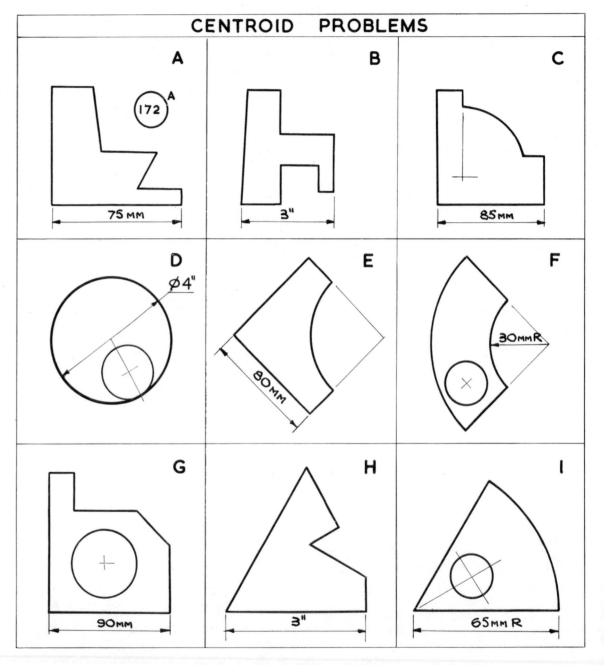

CENTROID PROBLEMS

A

172 A

75 MM

B

3"

C

85 MM

D

Ø4"

E

80 MM

F

30 MM R

G

90 MM

H

3"

I

65 MM R

MECHANISMS

Relative Velocity

Instantaneous Centre

Slider-crank

Four Bar Chain

Parallel Motions

Compound Motions

173. Relative Velocities The motion of the ends of a link may be shown by a vector diagram. The direction and magnitude of the given forces are indicated by vectors to scale and a triangle of forces completed. The magnitude and direction of the force at a second arm C may be found by drawing the link polygon. Notice that the line *ac* in the polygon is at right angles to AC, and similarly *cb* to CB. The line *oc* indicates the direction and magnitude of the force at C.

174. Instantaneous Centre In this second method consider the link A B in a circle centre I. If the dotted lines are drawn at right angles to the forces at A and B they intersect in centre I of the circle. The direction and magnitude of a force at C may be found by joining C to I and drawing the force line at right angles. The triangle of force is shown.

175. Slider and Crank Two methods of obtaining the direction and magnitude of the forces acting at points on the mechanism are shown. In the first, the forces are shown in the polygon. Draw the crank and slider to a suitable scale and to the given data. The force V*c* is always at right angles to the link CO and is indicated by r.p.m. or by radians per sec. The velocity of P the slider is indicated in ft/sec. To construct the polygon, draw *op* parallel to O P and to a suitable scale. Draw *oc* parallel to V*c*, draw *pc* parallel to a perpendicular to C P. The magnitude of V*c* is given to scale *co*. The force at D is indicated by V*d*, the position of *d* being obtained by transverse parallels from C D arranged on a line from *p* giving the same proportions as the division of the connecting rod.

176. Instantaneous Centre In this method, the centre I is found by drawing a perpendicular to PO, and extending the line of the link CO to intersect. The magnitudes of points D on the line PC may be found by joining the point to I, and its direction by drawing a perpendicular as shown in the diagram.

The velocity of A relative to O = the vector sum of velocity B.

Velocity of A relative to O = Vel. A + Vel. B, i.e.

$$oa = ob + ba.$$

The angular velocity of A B is

$$\frac{ab}{AB} \text{ radians per sec } \frac{\text{ft/sec}}{\text{L. link ft}}.$$

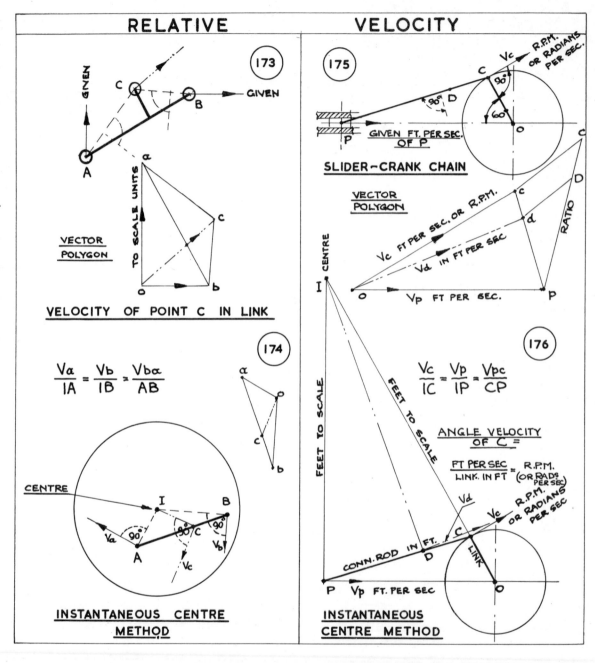

RELATIVE   VELOCITY

VECTOR POLYGON

VELOCITY OF POINT C IN LINK

$$\frac{Va}{IA} = \frac{Vb}{IB} = \frac{Vba}{AB}$$

CENTRE

INSTANTANEOUS CENTRE METHOD

GIVEN FT. PER SEC. OF P

SLIDER-CRANK CHAIN

VECTOR POLYGON

INSTANTANEOUS CENTRE METHOD

$$\frac{Vc}{IC} = \frac{Vp}{IP} = \frac{Vpc}{CP}$$

ANGLE VELOCITY OF C =

$$\frac{\text{FT PER SEC}}{\text{LINK. IN FT}} = \frac{\text{R.P.M.}}{\text{(OR RADS PER SEC)}}$$

$1 \text{ rev.} = 2\pi \text{ radians} = 360° \text{ (radian} = 57° \, 18').$

$$N \text{ r.p.m.} = \frac{N}{60} \text{ r. per sec}$$

$$= 2\pi \times \frac{N}{60} \text{ radians per sec}$$

$$= \frac{\pi}{30} N \text{ radians per sec}$$

also $\quad w \text{ radians per sec} = \frac{30}{\pi} \text{ r.p.m.}$

Relationship—linear velocity to angular velocity: length
of arc to angle in radians.

Velocity $\qquad$ linear $v = wr \qquad$ angular $w = v/r$
Acceleration $\quad$,, $\quad f = \propto r \qquad$,, $\quad \propto = f/r$

Angular Velocity:

One radian per sec $= 9\cdot5493$ r.p.m.
$\qquad\qquad\qquad = 57\cdot296°$ per sec.
One degree per sec $= 0\cdot16667$ r.p.m.
$\qquad\qquad\qquad = 0\cdot017453$ radians per sec.
One r.p.m. $\qquad = 6°$ per sec.
$\qquad\qquad\qquad = 0\cdot10472$ radians per sec.

177. Four Bar Chain In this mechanism, given the
velocity of D, the relative velocity of C may be found by
the instantaneous *ce* centre method, by extending the
lines of links A D and B C until they intersect. A point E
may be fixed at any position on D C and the force indicated by joining to I, the centre, and drawing a perpendicular to this line. The triangle of forces is also shown.

178, 179. Watts' Straight Line Motion Two diagrams
are given. In the first, equal radial links are joined by a
similar connecting link. The locus of a midpoint P on the
connecting link gives a straight line movement for much
of its travel. In the second case, unequal links are joined
by a connecting link, and the point P divides the connecting link inversely as the lengths of the radial links.
The locus of P is a straight line approximately for much
of its travel.

180. Watts' Quick Return crank and slotted plunger.
This mechanism is used to operate slides requiring a
quick return on the recovery stroke. The link is adjust-
able. The mechanism should be plotted through one
revolution.

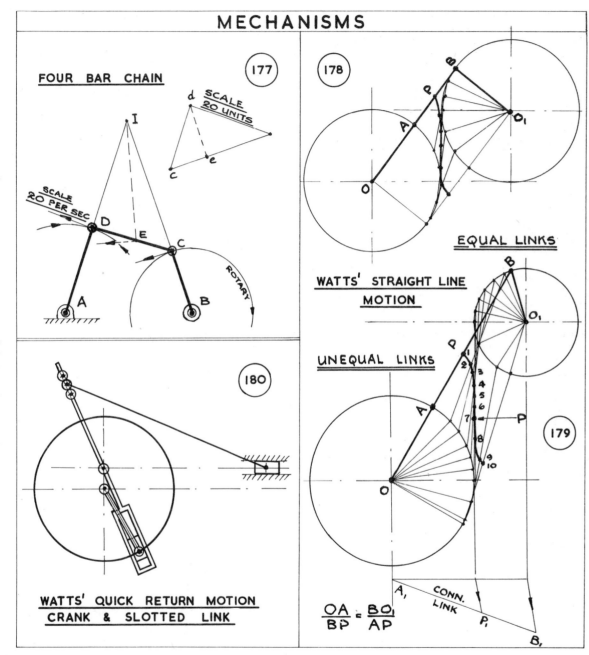

MECHANISMS

FOUR BAR CHAIN (177)

SCALE 20 UNITS

(178)

EQUAL LINKS

WATTS' STRAIGHT LINE MOTION

UNEQUAL LINKS

(179)

(180)

WATTS' QUICK RETURN MOTION
CRANK & SLOTTED LINK

$$\frac{OA}{BP} = \frac{BO_1}{AP}$$

181. Crank and Slotted Link The locus of an end point P is shown. The crank is drawn at twelve positions, and the position of the link positioned at these twelve stations. The curve can be drawn through the twelve plotted positions of P.

182. Linked Cranks The cranks and links should be plotted in twelve positions. Draw a fair curve through the positions of P and P_1. Notice that crank arms are rigid and can only describe arcs from their centres. Notice that links are rigid and pinned to the cranks; lengths measured along the length of the link are constant, in any position.

183. Slotted Link and Crank To enable the crank pin to work directly in the link, a suitable slot must be provided. Other links and slides can be attached. Plot the locus of a point P on the second link.

184. Toggle Action In the mechanism shown, the crank gives a quick S.H.M. clearing action for much of its stroke, exerting compression as required clamping action at the end of the stroke.

185. Cam and Plunger A simple circular cam actuates a spring loaded follower. This mechanism can be used to open a valve or a switch, or to operate a lever or link, one or more times per revolution of the cam. See cams.

186. Crank and Offset Linked Plunger The crank operates a pendulum link which is also linked to a pump ram. The line of the crank centre is offset from the point of attachment to the pendulum link, giving a quicker return action to the ram. Plot the mechanism in twelve positions and calculate the movement of the ram. Plot the locus of a midpoint P in the link connecting the crank pin to the pendulum link.

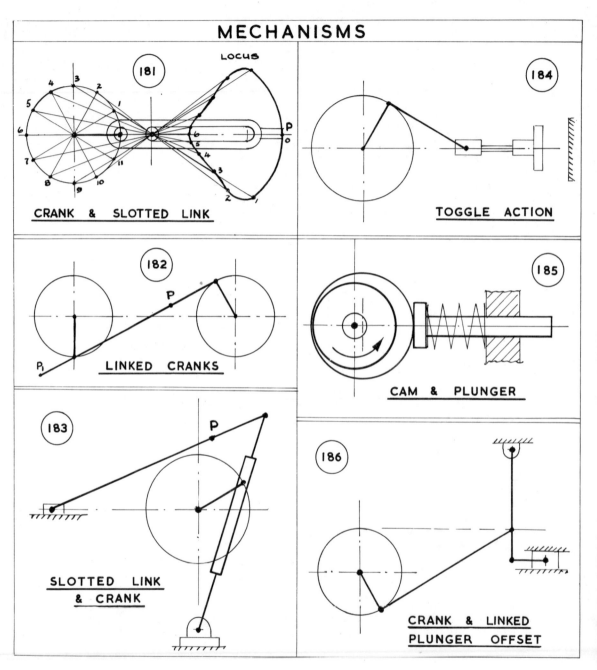

MECHANISMS

CRANK & SLOTTED LINK

LINKED CRANKS

SLOTTED LINK & CRANK

TOGGLE ACTION

CAM & PLUNGER

CRANK & LINKED PLUNGER OFFSET

88

187. Crank and Link A crank moves a link, the end of which moves on the inside of an arc. Plot the locus of any point P on the link. Substitute an elliptical curve for the arc, and compare the locus. Other curves may be substituted, parabolic, hyperbolic.

188. Radial Slotted Link The link swings from end A, and a slide C moves from A to B and back whilst the link swings through 90°. The diagram shows the locus of C. A second locus of the slide is shown. Plot the complete locus of C whilst the link rotates through 360°. Plot the locus of C when the link passes through 360° and the slide moves from A to B. Name the curve.

189. Bell Crank and Cam A cam giving S.H.M. moves a bell crank (two angled rigid arms) which has a swivel shoe attached. Plot the locus of A. Design a similar system to give $\frac{1}{2}''$ movement to the shoe, the ratio of the arms being 2:3.

190. Two Slotted Links and a Cam A S.H.M. cam works two slotted links which have a common slide C. Plot the locus of the centre of the slide. Vary the fulcrum points of the links. Substitute other cam forms shown on other pages.

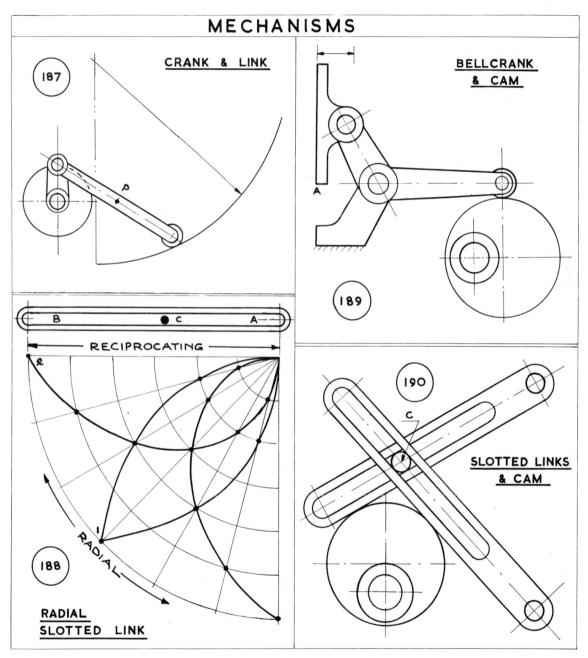

MECHANISMS

CRANK & LINK

187

P

BELLCRANK & CAM

A

189

B ● C A

RECIPROCATING

RADIAL

188

RADIAL SLOTTED LINK

190

C

SLOTTED LINKS & CAM

190A. Two Simple Harmonic Motions arranged at right angles. Two rotating cranks work in slotted links, the extensions of which combine to act upon a pin P. The speeds of the cranks are equal, though the length of the cranks may be unequal. To obtain the locus of P, divide the sweep circle of each crank into the usual twelve divisions and project the stations to give the points through which the locus may be drawn. Note that the locus is an ellipse.

190B. Unequal Speeds If two cranks turn at different speeds, the locus may be plotted as shown in the example. Two cranks of unequal length are arranged to work in slotted links at right angles (only the schematic diagram is shown). The speed ratio of the cranks is 2:1. The locus is plotted as shown by projecting from the twelve points on the slower crank circle, and six points (used twice) on the faster turning crank circle.

Exercises

1. Plot the locus of P when the cranks (in A) turn in the opposite direction (anti-clockwise).

2. Plot the locus when the two cranks start together at top dead-centre.

3. Plot the locus (in B) when the top crank starts at No. 1 position, and the lower crank starts at top dead-centre.

4. Plot the locus when the speeds of the two cranks are interchanged.

5. Plot the locus when the speeds of the two cranks are in the ratio of 3:1 (one revolution of the slower crank will be sufficient).

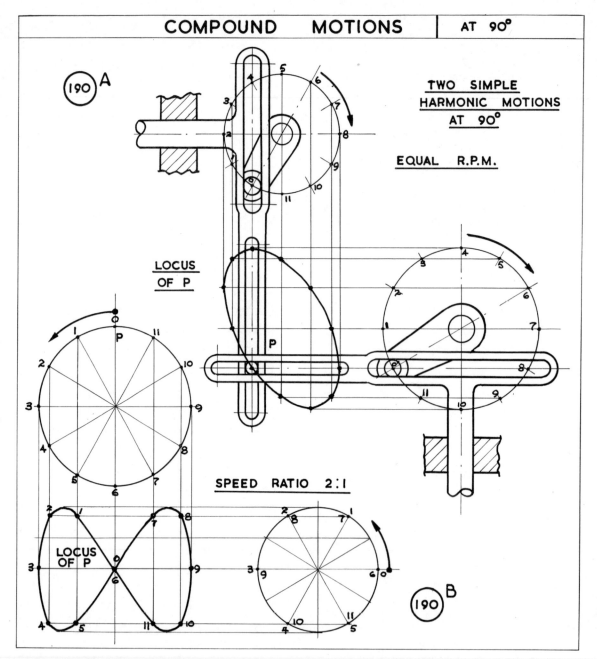

COMPOUND MOTIONS | AT 90°

TWO SIMPLE HARMONIC MOTIONS AT 90°

EQUAL R.P.M.

LOCUS OF P

SPEED RATIO 2:1

LOCUS OF P

INCLINED PLANE

True Lengths

Auxiliary Projection

OBLIQUE PLANE

Traces

Skew Lines

Dihedral Angle

Tangent Planes

INCLINED PLANE

191. Projections of a Line on a plane inclined either to the main horizontal or main vertical plane. The projections of a line and traces of the inclined plane are given; the isometric pictorial view should assist in envisualising the position (*a*) of the line itself, (*b*) the position of the plane. The true length of the line is obtained by rabatment. The line is treated as being part of the hypotenuse of a triangle which is rotated until it lies parallel to the main plane, when its true shape can be seen, and the true length of the line measured off. This system is known as 'triangulation' and is much used in obtaining developments; it is dealt with in a later section. Where an inclined plane cuts a solid, an auxiliary plan (for a plane inclined to H.P.), or an auxiliary elevation (for a plane inclined to V.P.) would have to be projected at right angles to the inclined plane to obtain the true shape (length of lines) of the section face. This is shown in the following diagrams.

192. Traces of a Line which intersect in the second angle. The isometric pictorial view should help to make this clear.

193. Traces of a Line which intersect in the fourth angle. Projections and an isometric view are given.

194. Section on an Inclined Plane A solid is shown cut by an inclined plane, the traces of which are given. This is an application of the plane to give a sectional view of the object. An auxiliary plan normal (at right angles) to the plane will give the true shape of the sectioned face of the object.

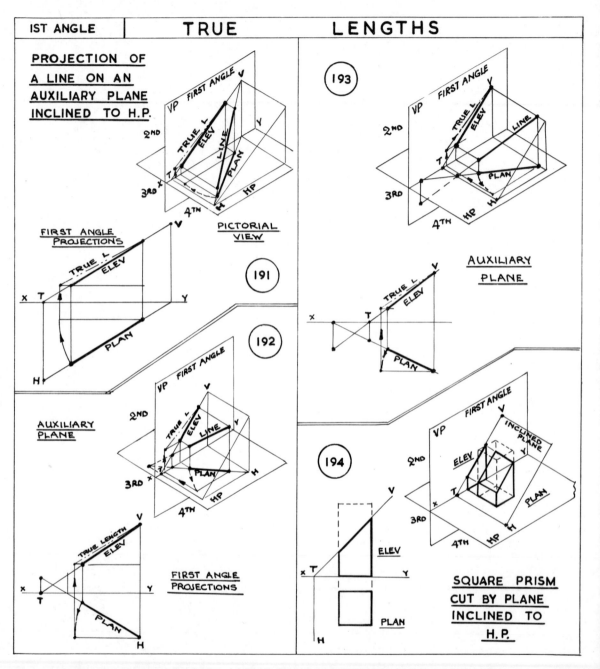

PROJECTION OF A LINE ON AN AUXILIARY PLANE INCLINED TO H.P.

FIRST ANGLE PROJECTIONS

PICTORIAL VIEW

191

192

AUXILIARY PLANE

FIRST ANGLE PROJECTIONS

193

AUXILIARY PLANE

194

SQUARE PRISM CUT BY PLANE INCLINED TO H.P.

92

195. Projections of a Line on a plane inclined to the main V.P. Traces of the inclined plane are given, V T H. The isometric pictorial view should help in understanding the position of the line and the plane. The true length of the line is obtained by rabating the line until it is parallel to the main V.P., projecting into the elevation, then joining to the original height line.

196. Square Prism Cut by Inclined Plane. The plane is inclined to the vertical plane and its traces are V T H. Simple projection of the points where the plane cuts the prism in the plan are projected to the elevation. The true shape of the sectioned surface would be found by projecting an auxiliary elevation on to the inclined plane. The method is shown in the next diagrams.

197. Auxiliary Plan shows true shape of the sectioned surface when a prism is cut by a plane inclined to the main horizontal plane. Widths remain the same when measured from the X Y line.

198. Auxiliary Elevation shows true shape of the sectioned face when a solid is cut by a plane inclined to the main vertical plane.

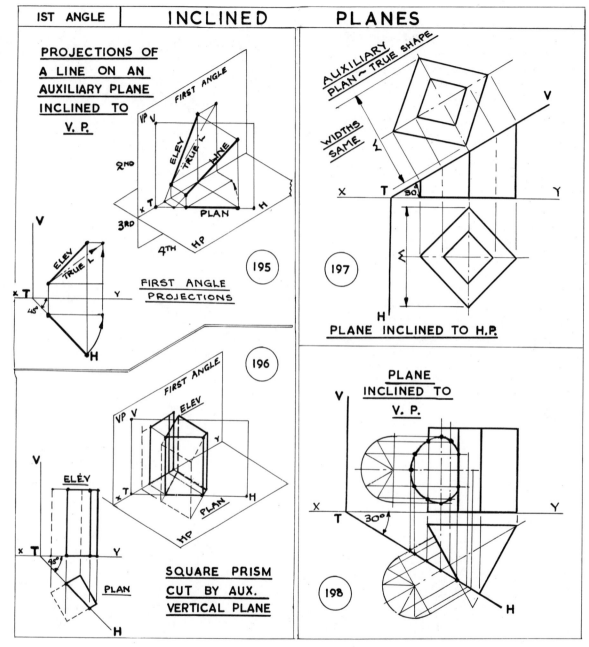

PROJECTIONS OF A LINE ON AN AUXILIARY PLANE INCLINED TO V. P.

FIRST ANGLE PROJECTIONS

195

196

SQUARE PRISM CUT BY AUX. VERTICAL PLANE

AUXILIARY PLAN ~ TRUE SHAPE

WIDTHS SAME

197

PLANE INCLINED TO H.P.

PLANE INCLINED TO V. P.

198

93

199. An Hexagonal Pyramid tilted at 30° to the main horizontal plane is cut by a plane inclined at 45° to the H.P. The true shape of the sectioned face is obtained by projecting an auxiliary plan.

200. A Cylinder and Prism are cut by a vertical plane which is inclined at 30° to the V.P. The true shape would be shown by an auxiliary elevation.

201. A Roof and Stack are cut by a plane inclined at 30° to the V.P. An auxiliary elevation gives the true shape of the sectioned face and length of lines thereon.

202. A Pierced Cylinder and Prism are cut by a vertical plane which is at 30° to the main V.P. An auxiliary elevation shows the true shape of the section.

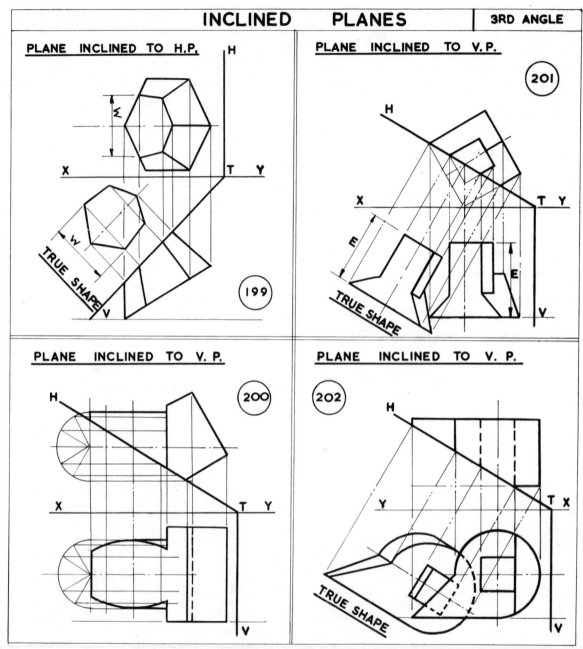

INCLINED   PLANES   3RD ANGLE

PLANE INCLINED TO H.P.

PLANE INCLINED TO V.P. (201)

(199) TRUE SHAPE

E E TRUE SHAPE

PLANE INCLINED TO V. P. (200)

(202) PLANE INCLINED TO V. P.

TRUE SHAPE

Exercises

203. Inclined Plane Scale up the diagrams three times size shown.

A. Obtain the true length of the line shown.

B. Obtain the true length of the lines shown. Show the traces of the planes.

C. The elevation and trace of a line are shown, find its true length.

D. Find the true length of the line, and the traces of the plane it lies in.

E. Obtain the true length of the line shown and indicate the traces of the plane.

F. The plan of a line and the traces of the plane it lies in are shown. Draw the elevation and find its true length.

G. An inclined plane, traces V T H, cuts a triangular pyramid. Draw the completed plan and elevation and project the true shape of the cut face.

H. A triangular prism is cut by an inclined plane V T H through point P. Draw the plan and elevation, and the true shape of the cut face.

I. The cone shown is cut by an inclined plane, traces V T H, which passes through point P. Draw the plan, elevation and the true shape of the cut.

J. A grooved rectangular prism is cut by an inclined plane V T H shown. Draw the projections. Project an auxiliary elevation to obtain the true shape of the cut face of the prism.

K. The tube shown is cut by a plane, traces V T H. Project the true shape of the cut face. Name the auxiliary view required.

L. The two interpenetrating cylinders are cut by an inclined plane, traces V T H. Draw the true shape of the cut face.

1. Work out the exercises as shown in First Angle projection.

2. Work out the exercises using Third Angle projection.

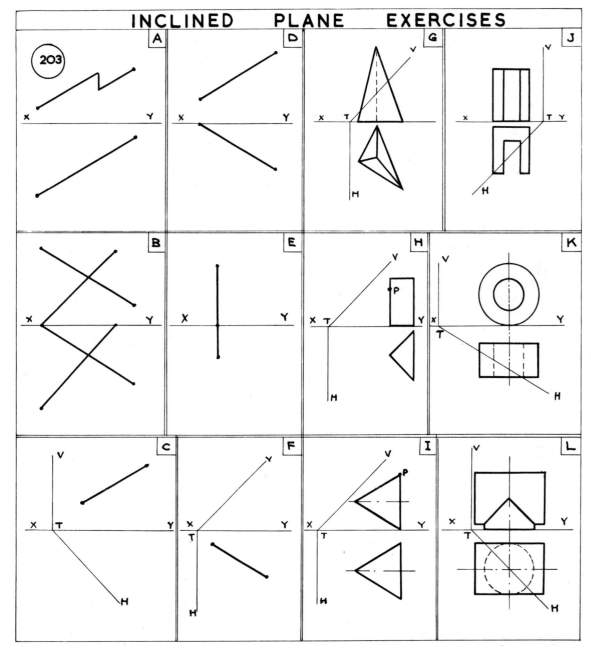

INCLINED PLANE EXERCISES

95

204. Auxiliary Projection Auxiliary plans and elevations projected on planes inclined to the principal vertical and horizontal planes of first and third angles are required when special details are not shown on the principal planes.

In the diagrams, a first auxiliary elevation is shown projected from the plan on a new x^1y^1 line. Heights remain the same in all elevations, and the projector from the plan is cut off to the same height as in the original elevation, as H.

Second Auxiliary Plan A second plan of the object may now be projected from the auxiliary elevation just completed.

A datum line x^2y^2 is drawn at the angle required or stated, and projectors from points in the first auxiliary elevation drawn. These are cut off to the same width measurements W below the datum line of corresponding points. The marked points on the projectors are joined to give the shape of the second auxiliary plan.

Notice that the same W measurement is used for two projector lines, since in the plan, the lower and upper points of a front edge are vertically above one another.

First Angle projection has been used in this diagram.

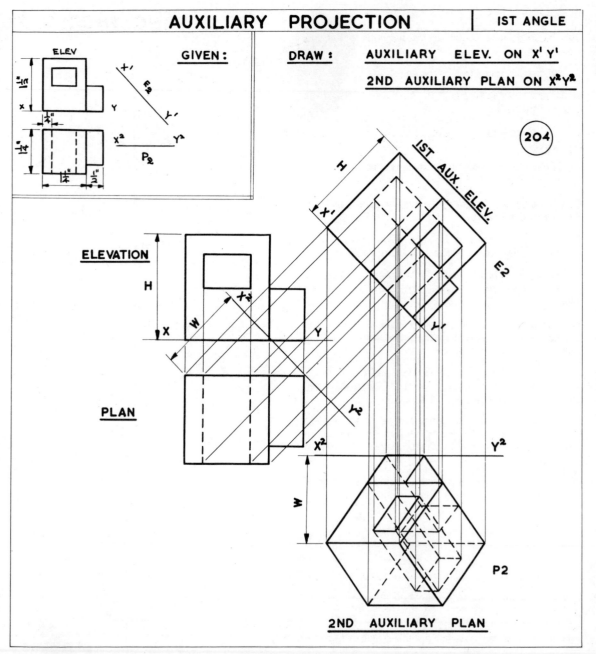

96

205. Auxiliary Projection In the problem given, third angle projection, a first auxiliary elevation is required, and a second auxiliary plan projected from the auxiliary elevation. The procedure is similar to that of the preceding page, first angle drawing.

Draw the new x^1y^1 line at the stated angle and in a convenient position. Project the salient points from the plan, and cut off to corresponding heights H as shown. Join the points to complete the first auxiliary elevation.

Second Auxiliary Plan Draw the datum line x^2y^2 and the new x^2y^2 line at the required angle and in position.

Project from the auxiliary elevation, and mark off the W measurements on the projectors.

Join the points to complete the second auxiliary plan.

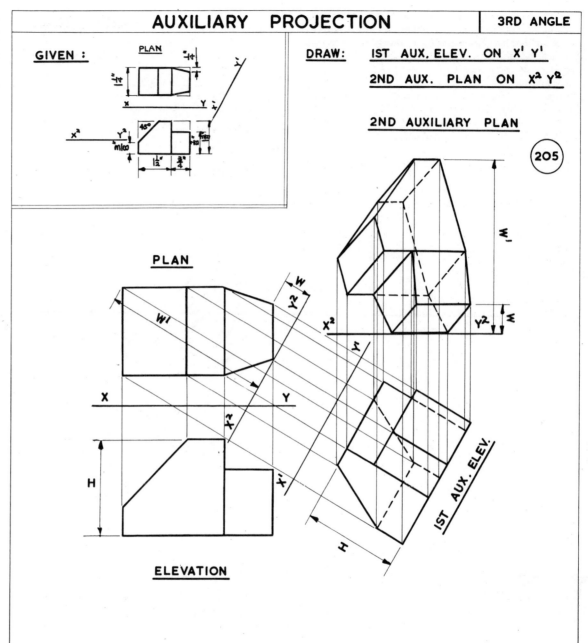

GIVEN :

DRAW: IST AUX. ELEV. ON X¹ Y¹

2ND AUX. PLAN ON X² Y²

2ND AUXILIARY PLAN

205

PLAN

ELEVATION

IST AUX. ELEV.

206. Auxiliary Projection A shaped bracket is given; draw in first angle, the original plan and elevation, f.s., project an auxiliary first plan. From this project a second auxiliary elevation. A further auxiliary elevation is required to show the shape of the hole in the bracket in that position.

Project the first auxiliary plan at the required angle from the original elevation, on x^1y^1.

Draw the datum line and new x^2y^2 line at the stated angle. Project the second auxiliary elevation from the first auxiliary plan.

The curve may be projected from centre line intersections, and four ordinate points on the compass arc shown in the construction on the original elevation.

When the examples shown have been understood, any of the objects shown for isometric exercises may be used for auxiliary projection. Common angles of the setsquare are usually asked for 30°, 60° and 45°. First and Third Angle projection should be used in successive drawings to gain practice.

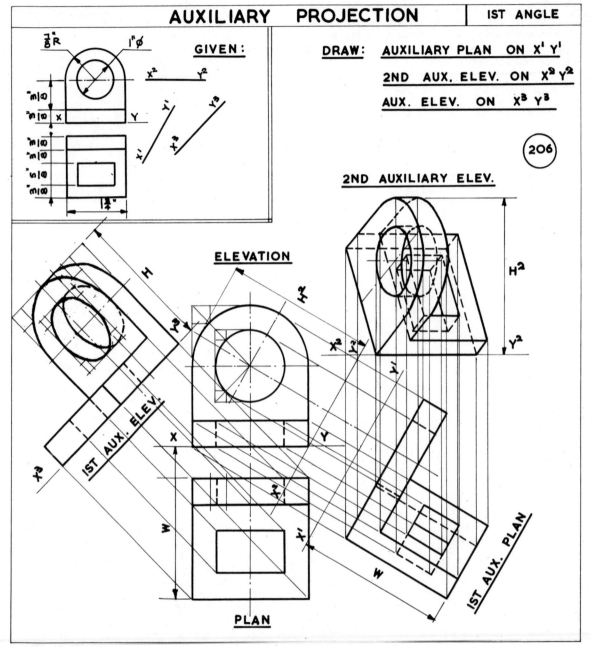

AUXILIARY PROJECTION — 1ST ANGLE

GIVEN:

DRAW: AUXILIARY PLAN ON X¹ Y¹
2ND AUX. ELEV. ON X² Y²
AUX. ELEV. ON X³ Y³

206

2ND AUXILIARY ELEV.

ELEVATION

PLAN

1ST AUX. ELEV.

1ST AUX. PLAN

207. The Oblique Plane The traces of an oblique plane and one projection of two points P and Q are given, in both first angle projection and third angle projection.

Notice that neither trace can be at right angles to the *xy* line.

208. First Angle Projection The traces VTH of the oblique plane and the projections of the two points P and Q on the main HP are given. Before the projection of P and Q can be made to the VP, their heights above the *xy* line must be obtained by first projecting an auxiliary elevation on x^1y^1 which is at right angles to the horizontal trace TH, and converts the oblique plane into an inclined plane. The height HV is the same for all elevations of point O. Points P and Q are now projected to the inclined plane and shown at P_1 and Q_1 giving their true heights above the *xy* line. These heights may now be measured on the projectors in the main vertical plane giving P_2 and Q_2 completing the required projection of the points to the main planes HP and VP.

209. Third Angle Projection The traces of an oblique plane HTV and one projection on the HP of two points P and Q are given. Complete the projections of the points to the vertical plane.

Draw the *xy* line and traces as given; place the points on the HP as given. Project an auxiliary elevation to convert the oblique plane into the inclined plane, height VH at point O. Project lines from P and Q to cut the line of the inclined plane in P_1 and Q_1. Transfer these heights to the main VP to obtain P_2 and Q_2.

The two methods are similar in procedure, though the position of the main planes differ.

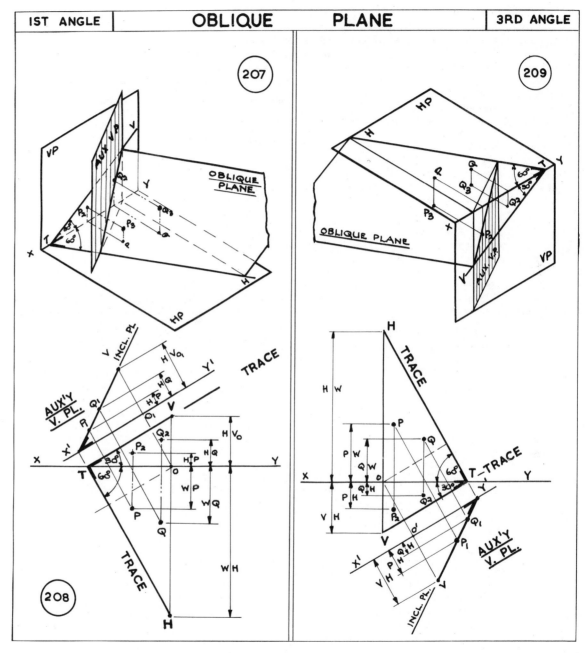

210. A. Projections of a Line on an Oblique Plane. The traces of an oblique plane V T H are shown. By means of an auxiliary elevation or plan projected in line of the appropriate trace the oblique plane is changed into an inclined plane. Lengths projected to the inclined plane are then transferred to a second auxiliary plan or elevation to give true lengths.

A line A B is given in plan. Project an auxiliary elevation in line with T H and on x^1y^1. This gives the line on an inclined plane, and heights H^1 and H^2 which can now be transferred to the original elevation. Projectors from the plan intersecting the height lines give the elevation of the line. A further elevation on x^2y^2 projected in line with trace V T intersecting with the height lines gives the true length of the line.

211. B. The elevation of a line lying on an oblique plane traces V T H is shown. Project an auxiliary plan. Width from the X Y line remains the same, and the line now appears lying on the inclined plane. W^1 and W^2 intersect projectors from the elevation and the plan of the line established. Develop the surface of the inclined plane, transfer the height lines to this plane, and the true length of the line is found.

The isometric sketches in the following diagrams should help in understanding how the oblique plane is transformed into an inclined plane by projecting the auxiliary elevations or plans.

212. An Oblique Plane is inclined to both horizontal and vertical main planes, and its traces appear as indicated V T H.

The traces are the lines of its intersection with the main vertical and horizontal planes.

213. The True Angles of Inclination to the main planes may be obtained by projecting the point of intersection of a perpendicular to V T and T H to the xy line and joining this to V and H respectively. This is the rabatment of a triangle taken normal to the oblique plane and main planes. The isometric diagram should help to make this clear.

214. The True Angles may also be found by projecting auxiliary elevations and plans in line with the traces as shown in the middle diagram. Notice that heights remain the same in all the elevations, and widths below the xy lines remain the same in all the plans, of associated projections.

215. Given a Triangle projected on an oblique plane, the traces of which are also given, find the true shape.

Project an auxiliary elevation on the line of T H. The points a^1, b^1, c^1 and L_1, L_2, L_3 are true lengths on the inclined plane. The points are now brought to a position where the widths may be projected lengths L_1, L_2, L_3. This is shown in the true shape diagram.

216. Given the Position of Point Q and the traces of an Oblique Plane V T H, project the plan. Draw the traces and the given point Q in elevation. Project an auxiliary elevation in line with T H. H_2 remains the same height in all elevations, and enables the inclined plane (end view of the oblique plane) to be drawn. Transfer H_1 to the auxiliary elevation, the intersection gives Q^1. Project this point back to the plan to intersect a vertical projector from Q, resulting in Q^2, the required plan.

The dotted line in the auxiliary elevation is the true distance of Q from the H P.

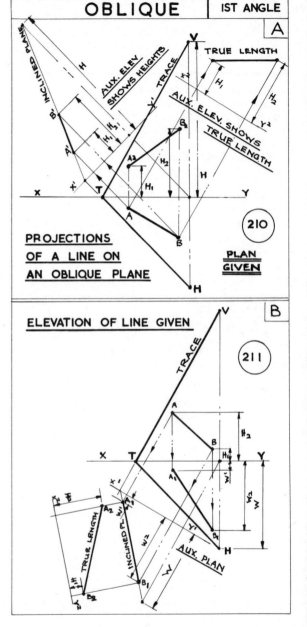

OBLIQUE · IST ANGLE

A

PROJECTIONS OF A LINE ON AN OBLIQUE PLANE

210

PLAN GIVEN

B

ELEVATION OF LINE GIVEN

211

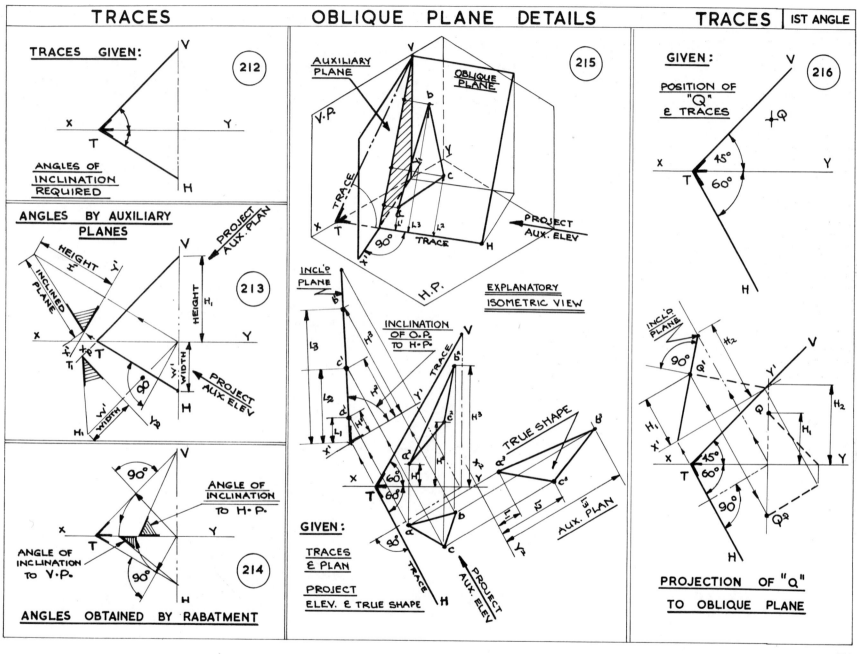

TRACES GIVEN: 212

ANGLES OF INCLINATION REQUIRED

ANGLES BY AUXILIARY PLANES 213

PROJECT AUX. PLAN

HEIGHT H'

Y'

INCLINED PLANE

HEIGHT H₁

HEIGHT

PROJECT AUX. ELEV

WIDTH W'

90°

W' WIDTH

Y₂

H₁

ANGLES OBTAINED BY RABATMENT 214

90°

ANGLE OF INCLINATION TO H·P·

ANGLE OF INCLINATION TO V·P·

90°

AUXILIARY PLANE

OBLIQUE PLANE 215

V·P·

TRACE

90°

TRACE

H·P·

EXPLANATORY ISOMETRIC VIEW

INCL'D PLANE

INCLINATION OF O.P. TO H·P·

L3

L2

L1

TRACE

H³

H²

H⁴

60°

60°

90°

TRUE SHAPE

AUX. PLAN

GIVEN:

TRACES & PLAN

PROJECT ELEV. & TRUE SHAPE

PROJECT AUX. ELEV

GIVEN: 216

POSITION OF "Q" & TRACES

45°

60°

INCL'D PLANE

90°

45°

60°

90°

H₂

H₁

PROJECTION OF "Q" TO OBLIQUE PLANE

217. Two Skew Lines, shortest distance between two straight lines neither parallel nor intersecting, one common perpendicular (shortest distance between), said to be at right angles when their projections on any plane normal to their common perpendicular are at right angles.

Draw the given lines and the *xy* line.

Draw a construction line E F parallel to C D and $E_1 F_1$ parallel to $C^1 D^1$ and obtain the intersections.

Extend and project A B and $A^1 B^1$ to give further intersections. Join the intersection points to give the traces V T and T H. Project an auxiliary plane on $x^1 y^1$ which is at right angles to T H. This converts the oblique plane into an inclined plane, and the lines A B and C D can now be projected from the plan to the inclined plane. The common perpendicular may now be drawn $O_3 P_3$, and these points projected back to the plan, giving O and P. Simple projection of O, P to the main vertical plane give O^1 and P^1 and completes the elevation.

$O_3 P_3$ is the length of the common perpendicular and the shortest distance between the two skew lines.

The isometric sketch should help in visualising the oblique plane and the skew lines.

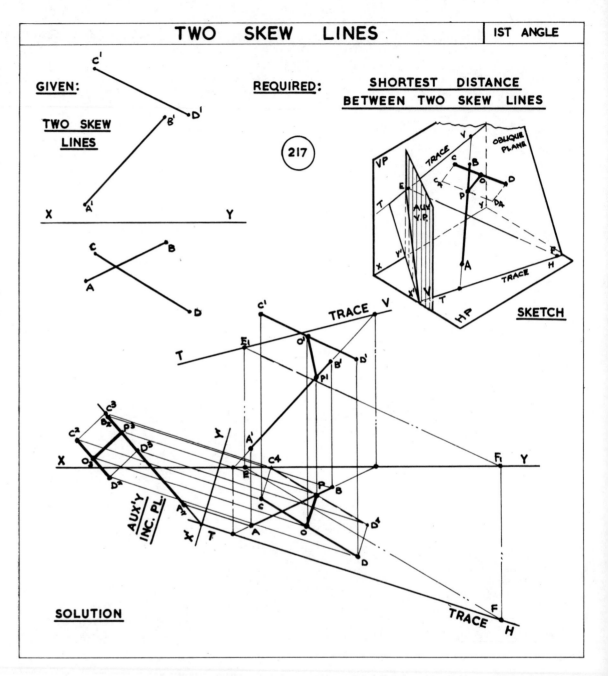

GIVEN:

TWO SKEW LINES

217

REQUIRED: SHORTEST DISTANCE BETWEEN TWO SKEW LINES

SKETCH

SOLUTION

218. Solids on an Oblique Plane First Angle Projection. Given the traces of an oblique plane VTH, and a cone standing on it, point P of the cone base circle touching the H P.

Draw the traces of the oblique plane and the *xy* line, in first angle projection.

Project an auxiliary elevation converting the oblique plane into an inclined plane, on a new x^1y^1 line. Draw the cone resting on the inclined plane.

Fix P by measurement on T H, and draw the ellipse in plan by projecting from the auxiliary elevation; the major axis of the ellipse will be the undiminished diameter of the original base circle, $1\frac{3}{4}''$. The apex of the cone is obtained by simple projection from the auxiliary elevation.

The elevation of the cone on the main VP is constructed by projecting points on the plan ellipse to the VP, and transferring the respective heights, H^2, H^3 (two), H^4 (three), H^5 (two) and P (zero). The base circle of the cone appears as an ellipse also in the vertical plane.

219. Pentagonal Pyramid on an oblique plane. Third Angle. Draw the traces. Project the inclined plane in line with H T. Draw the pyramid resting on the inclined plane. Project the base shape and apex back to the H P. Complete this main plan view.

Project the main elevation from points in the plan, transferring heights from the auxiliary elevation to their respective projectors on the main vertical plane.

Note that the method is the same in both cases, the object is drawn on the inclined plane which is the oblique plane viewed 'edge-on'.

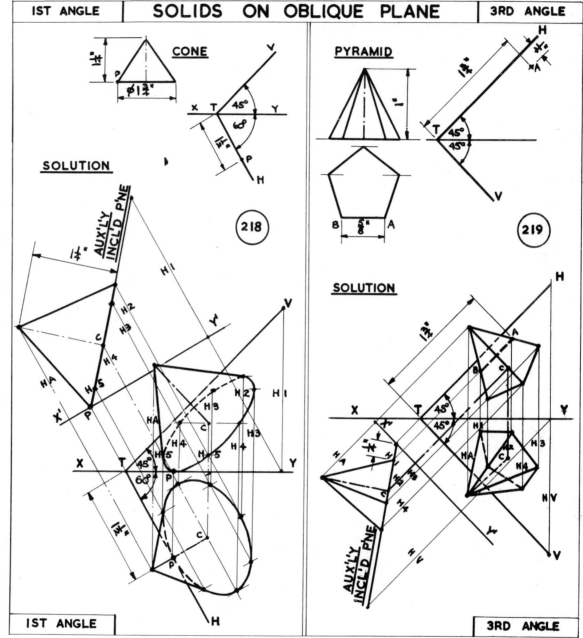

SOLIDS ON OBLIQUE PLANE

1ST ANGLE   3RD ANGLE

CONE

PYRAMID

SOLUTION

218

219

1ST ANGLE   3RD ANGLE

103

220. Oblique Section of a Triangular Prism, given the traces and the position of the prism. Draw the traces and the plan of the prism. Project an auxiliary elevation on x^1y^1, project *abc* to the inclined plane. Project an auxiliary plane below x^2y^2, taking widths from the original plan. This gives the true shape of the oblique section. The isometric view should help to make the projection clear.

221. Oblique Section of a Square Pyramid, given the traces and the position of the pyramid. Draw the traces and the outlines of the pyramid. Project an auxiliary elevation on x^1y^1. The line of the inclined plane gives heights of *abcd* which can now be transferred back to the original plan, and thence projected to the elevation which in turn gives the orthographic projections of the section. An auxiliary plan may now be projected on x^2y^2, which yields the true shape of the section. The isometric view should help in understanding the projections.

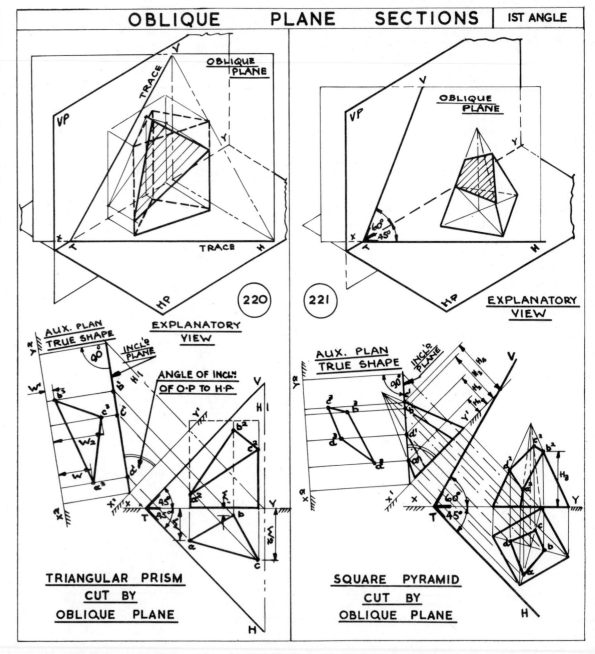

OBLIQUE PLANE SECTIONS | 1ST ANGLE

EXPLANATORY VIEW

TRIANGULAR PRISM CUT BY OBLIQUE PLANE

SQUARE PYRAMID CUT BY OBLIQUE PLANE

222. Oblique Section of a Cylinder given the traces and position of the cylinder. Draw the traces and simple orthographic views of the cylinder. Project an auxiliary elevation on x^1y^1. The line of the inclined plane cuts the cylinder giving heights which can be transferred via the plane to the original elevation. A fair curve drawn through the points completes the elevation. The true shape of the section may be projected below x^2y^2, widths being taken from the original plan.

223. Oblique Section of a Cone, given the traces and position of the cone. Notice in this case the angle between the traces is obtuse, and the isometric sketch should help to show the direction of the plane which rises to the observer from the xy line. Draw the traces and the simple lines of the cone. Project an auxiliary elevation on x^1y^1. The line of the inclined plane cuts the generators of the cone and thus yields heights. These can be used in the original elevation being projected back through the plan.

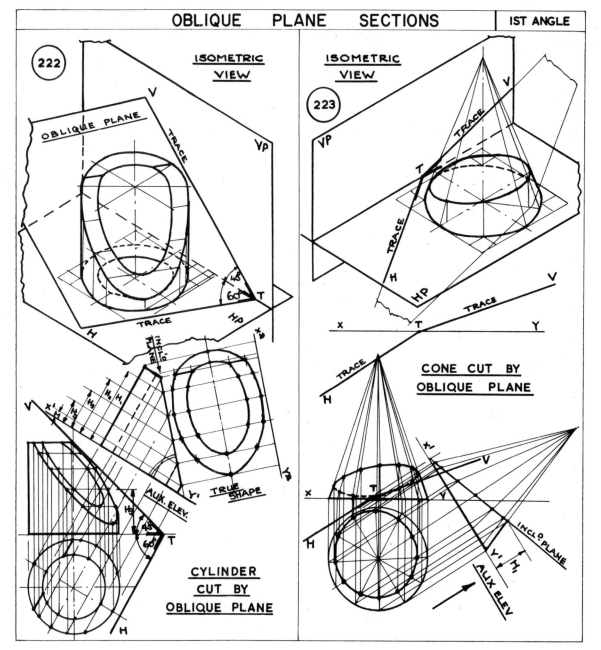

ISOMETRIC VIEW

ISOMETRIC VIEW

CYLINDER CUT BY OBLIQUE PLANE

CONE CUT BY OBLIQUE PLANE

Oblique Plane Problems

224. Given the Projections of a line A B and the traces of an Oblique Plane V T H, find the point of intersection of the line with the plane. Draw the *xy* line, projections of the line *ab*, and a^1b^1 and traces of the plane V T H as shown. Project an auxiliary elevation on x^1y^1 which is at right angles to T H. Convert the oblique plane into an inclined plane. Project the given line also to the auxiliary elevation from the plan to give a^2b^2, their heights being taken from the original elevation. The intersection of the line a^2b^2 and the inclined plane is the point P, which may now be projected back to the plan P_1 and thence to the elevation P_2.

225. Dihedral (True) Angle between two oblique planes. Draw the traces as given. Project the points *a* and *b* which are the projections of the line of intersection between the two planes. Project an auxiliary elevation on x^1y^1 which is parallel to *ac*. Draw *de* normal to the inclined plane, rabat to the plan as shown to give the true shape which shows the true angle between the two oblique planes. See also 227.

226. To Find the Traces of an Oblique Plane Draw the given plan and elevation of the laminae, extend *ab* and *ad*, a^1b^1 and a^1d^1 to give, by rabatment, intersection points 1 and 2 through which T H may be drawn. Find 3 and 4 by projection from *cd*, parallel to T H to cut the *xy* line, thence to intersect horizontal projectors from c^1 and d^1, through which V T may be drawn. The auxiliary elevation shown may be drawn as in previous problems to verify heights a^2, b^2, c^2, d^2.

This problem may now be extended as in Nos. 220 and 221 to find the true shape of the laminae when required.

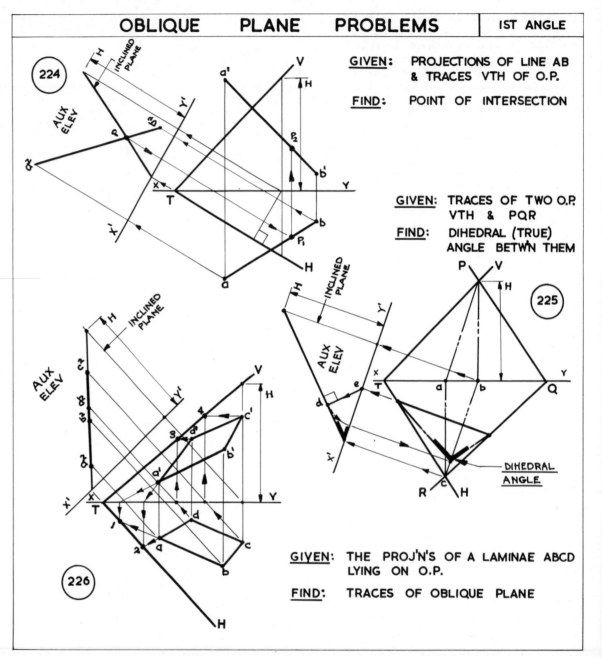

OBLIQUE PLANE PROBLEMS | 1ST ANGLE

224

GIVEN: PROJECTIONS OF LINE AB & TRACES VTH OF O.P.

FIND: POINT OF INTERSECTION

GIVEN: TRACES OF TWO O.P. VTH & PQR

FIND: DIHEDRAL (TRUE) ANGLE BETW'N THEM

225

DIHEDRAL ANGLE

GIVEN: THE PROJ'N'S OF A LAMINAE ABCD LYING ON O.P.

FIND: TRACES OF OBLIQUE PLANE

226

227. Dihedral Angle The dihedral angle between two planes is shown on a plane which cuts the two planes when normal to both.

Square Pyramid 1. Draw the plan and elevation of the solid. Project an auxiliary elevation as shown on x^1y^1. Draw in a plane shown by the line *ab*, which is at right angles to the projected angle line where two surfaces of the pyramid meet.

Rabat point *b* to the plan, join to the two corners of the base square. The dihedral angle is shown lined in, and represents the true angle between the two adjacent sides of the pyramid.

Triangular pyramid 2. In the projections of the triangular pyramid shown, the cutting plane *ab* can be drawn in immediately as angle between the two meeting sides is shown in true elevation. Point *b* can be rabatted to the plan and joined to the two corners of the base triangle giving the dihedral angle as shown.

The diagram also shows the method applied in Fig. 1., using an auxiliary plane enabling point *c* to be rabatted to the plan.

Hexagonal Pyramid 3. Draw the projections of the pyramid as shown. Draw the auxiliary plane, and the cutting plane. Rabat to the plan and join to the two corners as in the diagrams. Note that the cutting plane affects only two adjacent surfaces shown, and is not taken to the centre point of the base.

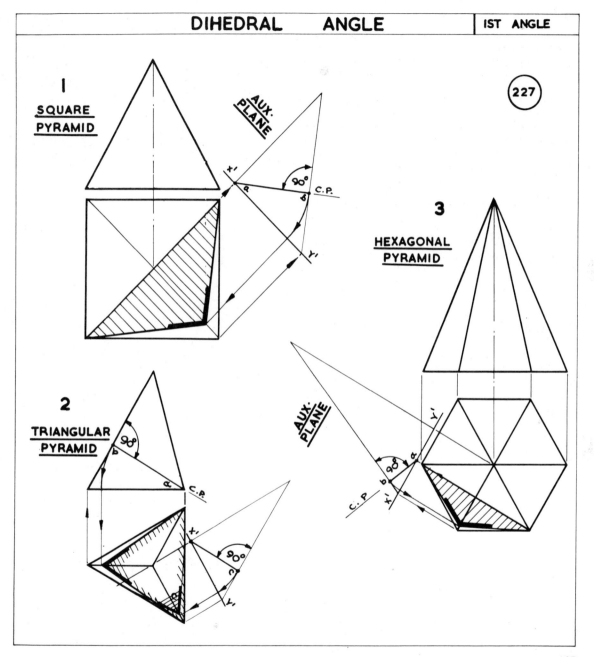

107

228. Tangent Planes Given the plan of a sphere and a point P, find the traces of the plane tangential to the sphere at the point.

Draw the plan and point given, project the elevation. P will lie on a section plane shown in the plan, and P_1 is projected as shown.

Project an auxiliary plan on x^1y^1 which is parallel to P_1O_1. P_2 is obtained by projecting from P_1, and the inclined plane is tangential to the circle at P_2. VT can now be drawn parallel to P_1P_2, and TH is drawn from T and at right angles to PO.

The auxiliary elevation completes the projection, the true angles of the plane to the HP and the VP are shown in the auxiliary elevation and plan. An alternative method is shown by the dotted lines. Draw the plan and elevation. Draw P_1a at right angles to P_1O_1 to give a on xy. Find b by simple intersection as shown. Pc is at right angles to PO. Find d by simple projection. Draw VT through d and parallel to P_1a. Draw TH through b and parallel to Pc.

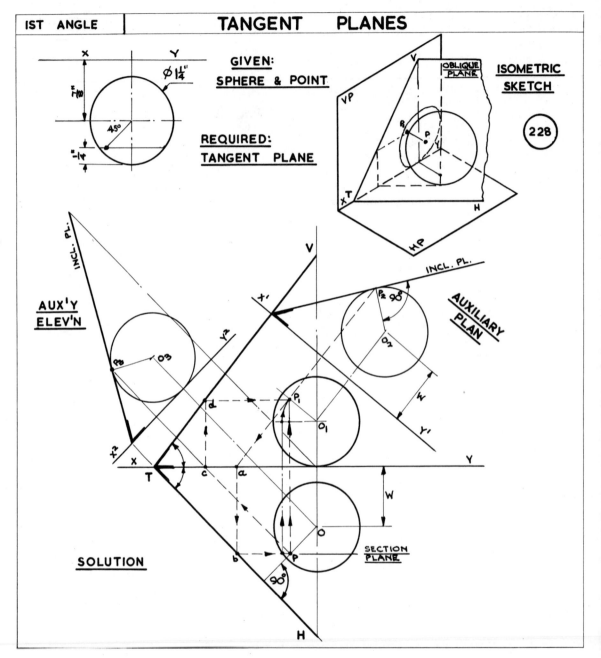

IST ANGLE

TANGENT PLANES

GIVEN:
SPHERE & POINT

REQUIRED:
TANGENT PLANE

ISOMETRIC SKETCH

228

AUX'Y ELEV'N

AUXILIARY PLAN

SOLUTION

SECTION PLANE

229. Tangent Plane to Cone   Given point P on the plan of a cone, find the traces of an oblique plane tangential at the point P.

Draw the plan and point P. Project the elevation of the cone. Draw a generator in plan passing through P; project to the elevation to obtain P_1.

Draw TH tangential to the base circle and at right angles to PO. Draw O*a* parallel to TH. Find *b* by simple projection. Join T to *b* to give trace VT.

Project an auxiliary elevation on x^1y^1 to give the true angle of inclination of the oblique plane to the HP.

In the case shown, the oblique plane lies in contact on the cone surface on the generator which includes P.

230. Tangent Plane to Cylinder   Given the one projection on the HP of a point P on a cylinder inclined to the VP, find the traces of the oblique plane touching the cylinder at the point.

Draw the plan and point P. Project an auxiliary elevation on x^1y^1 which is at right angles to the axis of the cylinder. Draw a projector from P parallel to the axis to give P_1 on the circle. Join P_1 to O, and draw the line of the inclined plane tangential at P_1. TH may now be drawn by projection from x^1y^1, and is parallel to the axis. The height of P_1 is shown in the auxiliary elevation, and can be transferred to the main VP to fix the position of P_2. V is found by drawing a perpendicular from *c* and is height HV transferred from the auxiliary elevation.

Alternatively, simple projection will give points *abc*, VT being drawn by joining T*b* and extending to V.

In this case, it will be seen that the oblique plane rests on the cylinder giving a line of contact including point P, the line being parallel to the axis.

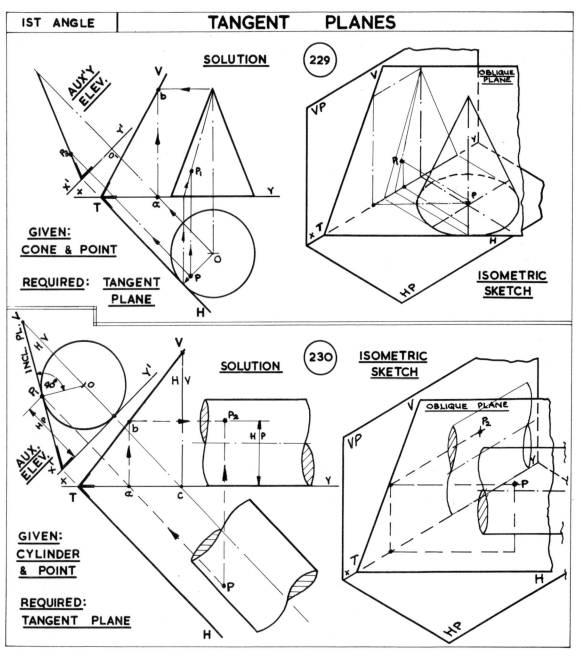

109

231. True Shape of a Triangle or of any plane area. The true shape of a plane figure may be found by the following method if the plan and elevation are given.

Draw the given elevation *abc* and project the plan $a^1b^1c^1$. Draw a constructional horizontal line *cd* in the elevation, and project to the plan to give c^1d^1.

Project an Auxiliary Elevation on x^1y^1 normal (at right angles to) line c^1d^1, to give a^2, b^2, c^2. Heights of these points are the same as in the original elevation from *xy*. Project an Auxiliary Plan on line x^2y^2, which is parallel to $a^2b^2c^2$, to give $a^3b^3c^3$. Width distances of these points are taken from x^1y^1 to the original plan as shown in the diagrams.

The true shape of the triangle is given in the auxiliary plan as indicated.

Refer also to the chapter on auxiliary plans and elevations first and second, given earlier.

The true shape of any figure may be found by the above method if the limit points of the horizontal construction line are projected with sufficient boundary points of the figure.

The true shape of the triangle could also be found by finding the true length of each side in turn by rabatment —see triangulation—and thence drawing the triangle from the true length of the sides.

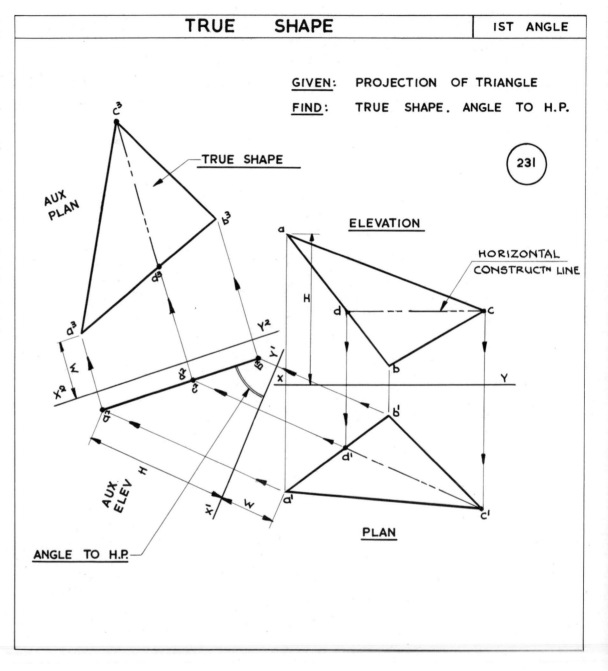

GIVEN: PROJECTION OF TRIANGLE

FIND: TRUE SHAPE. ANGLE TO H.P.

231

TRUE SHAPE

AUX PLAN

ELEVATION

HORIZONTAL CONSTRUCTⁿ LINE

AUX ELEV

ANGLE TO H.P.

PLAN

110

232. Line of Intersection between two planes. Where two planes intersect, given the projections, plan and elevation, the line of intersection may be found by the use of the horizontal construction line (as in the preceding diagram), by auxiliary elevations and second plan (see 204–206) which will give the 'edge view' of the two planes and consequently the true or dihedral angle. (See also Dihedral Angle diagrams, No. 227.)

Draw the given elevation and plan with the horizontal construction line *ah* as shown. Project an Auxiliary Elevation on x^1y^1 normal to a^1h^1, to give the points of intersection of the two planes, one of which, $a^2b^2c^2$ is seen in 'edge view'. Project the points of intersection of the planes back to the original plan and thence to the original elevation. The line of intersection between the two planes may now be drawn as shown in the diagrams.

The True Angle or Dihedral Angle between the planes may now be found by first projecting a second Auxiliary Elevation on line x^2y^2 as shown, and projecting from this view, a second Auxiliary Plan on line x^3y^3 normal to the line of intersection in the last view. This results in both the planes being shown in 'edge view'. and the true angles between the planes shown, as indicated in the diagram.

The true shape of each of the triangles can be found as a further exercise using the same method as shown in the previous page.

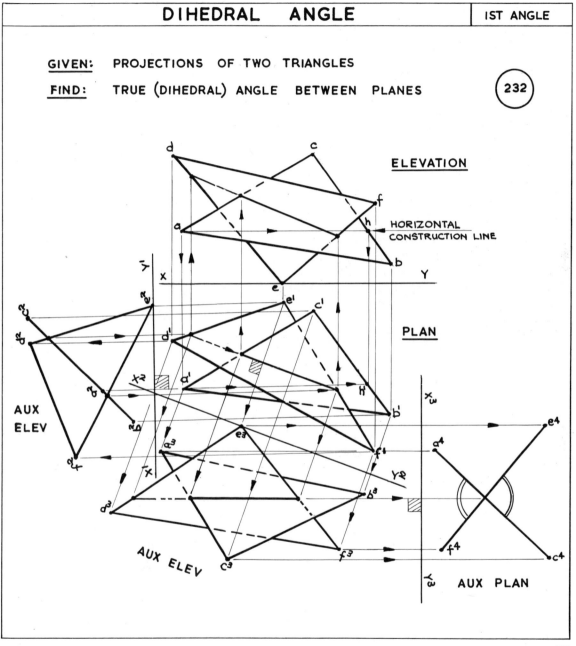

DIHEDRAL ANGLE — 1ST ANGLE

GIVEN: PROJECTIONS OF TWO TRIANGLES

FIND: TRUE (DIHEDRAL) ANGLE BETWEEN PLANES

232

ELEVATION

HORIZONTAL CONSTRUCTION LINE

PLAN

AUX ELEV

AUX ELEV

AUX PLAN

Exercises

232A. Inclined and Oblique Planes

A. Details of a heptagonal pyramid are shown; find the dihedral angle between two adjacent sides. Draw a 4″ diam. circle and construct the heptagon therein, see No. 12. This forms the plan; project the elevation. Project an auxiliary elevation, and proceed as in No. 227.

B. A segmental section is shown, cut by an inclined plane HTV, third angle projection. Draw the given views to f.s. scale, project the intersection points from the elevation to the plan. Project an auxiliary plan to show the true shape. See Nos. 196 to 202.

C. A cylinder is cut to give a lunar section. Traces of an oblique plane VTH, first angle projection are shown. Draw the cylinder as cut by the oblique plane passing through point P. Project an auxiliary elevation from the given plan; in this case, XV may be used as the new *xy* line as the angle is 45°. Proceed as in No. 212.

D. Third angle projections are given of a pierced shaped block and traces of an oblique plane which cuts the block at point P. See Nos. 208 and 220 to 226.

The simpler shaped blocks to be used as exercises in pictorial drawing may also be used as subjects for inclined plane and oblique plane exercises if further experience is sought. See Nos. 39 and 40.

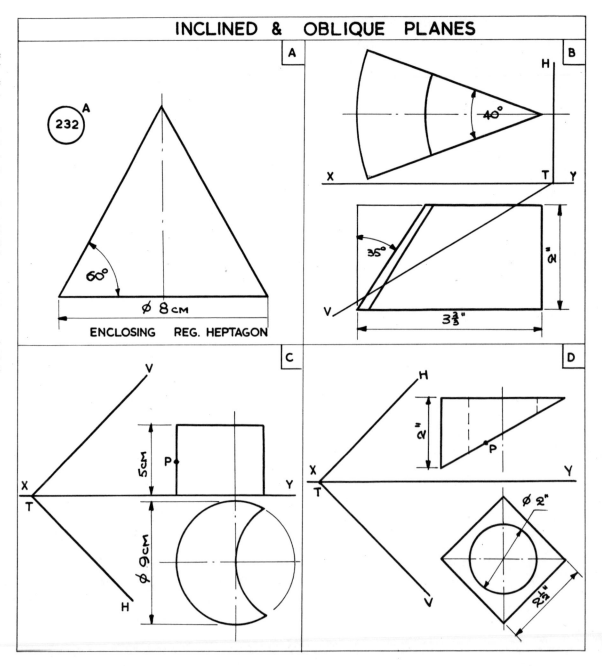

INCLINED & OBLIQUE PLANES

A
232
ENCLOSING REG. HEPTAGON
⌀ 8 cm
60°

B
H
40°
X T Y
35°
2″
3⅔″
V

C
V
X T Y
5cm
P
⌀ 9cm
H

D
H
2″
P
X T Y
⌀ 2″
2⅓″
V

112

SIMPLE DEVELOPMENTS AND INTERPENETRATION

Rectilinear Cylinders

Pyramids

Cones

Spheres

Palmate

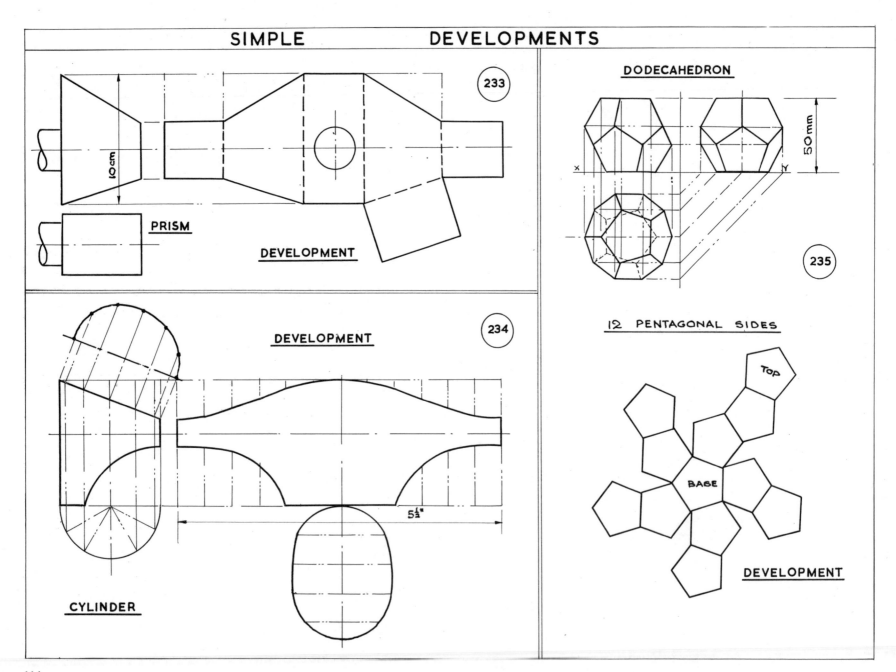

233

PRISM

DEVELOPMENT

10 cm

234

DEVELOPMENT

CYLINDER

5½"

DODECAHEDRON

50 mm

235

12 PENTAGONAL SIDES

TOP

BASE

DEVELOPMENT

Simple Developments

233. Rectangular Prism Simple projection of the faces in line. True lengths are shown either in plan or elevation and require no rabatment.

234. Cut Cylinder Intersection of the twelve generators with the lines of the cuts give points which may be projected to the development. The top face would be an ellipse. The lower face would be part circle.

235. Dodecahedron A solid having twelve regular pentagonal faces. The projections and development are shown.

236. Tetrahedron A solid having four equilateral triangular faces.

237. Octahedron A solid having eight equilateral triangular faces.

238. Hood Rectangular section. The development of the front face is shown, widths being taken from the elevation, true length of the line 0 to 9 being stepped off by dividers.

239. Sphere The development of the surface of a sphere may be approximated by drawing lines, whose length is πR. The curved shape is obtained by ordinates from sections of the hemisphere as shown.

240. Canister Rectangular section. Part development of an end is shown, widths being obtained from the plan, 0 to 13 being stepped off by dividers.

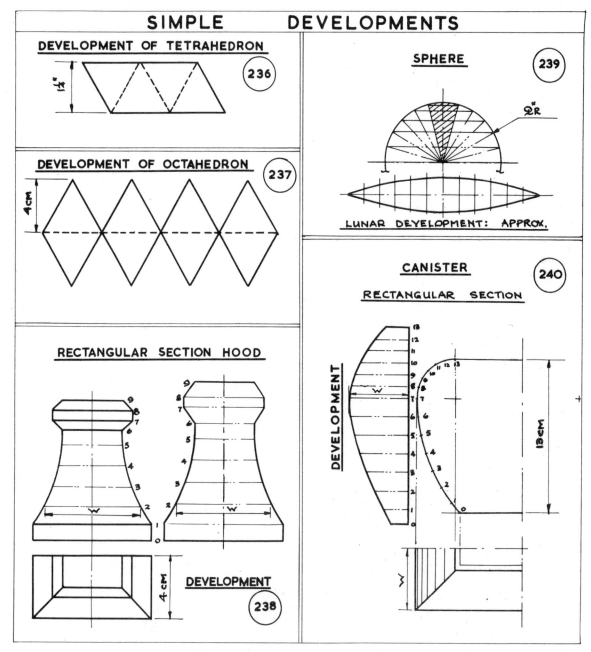

SIMPLE DEVELOPMENTS

DEVELOPMENT OF TETRAHEDRON 236

DEVELOPMENT OF OCTAHEDRON 237

RECTANGULAR SECTION HOOD

DEVELOPMENT 238

SPHERE 239

LUNAR DEVELOPMENT: APPROX.

CANISTER 240

RECTANGULAR SECTION

DEVELOPMENT

241. Interpenetration of Cylinders Equal diameters. Tee joint. Draw the plane and elevation. Divide the circles into twelve parts and project generators. The intersection of the appropriate generators give points on the line of interpenetration, in this case a straight line. The developments of the cylinders are best made in line with the portion developed. Twelve chordal distances have been taken to represent the unrolled cylinder, but since the developments are symmetrical, only three points need projecting.

242. Right-angled Junction Equal diameters. Only the elevation need be drawn, a construction semi-circle gives the generators. The line of interpenetration is a straight line. The development is symmetrical, and only a half development need be drawn, the other half being identical.

Greater accuracy may be obtained by drawing twice as many generators and by calculating the circumference as πD, but the twelve generators are usually accepted.

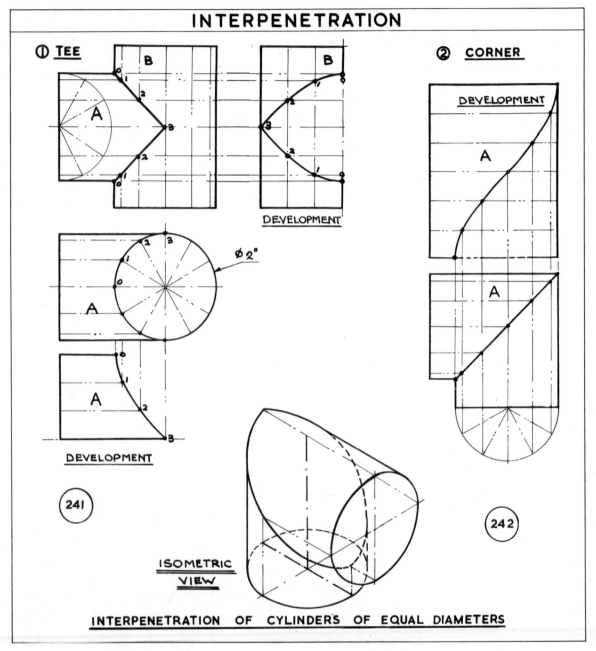

INTERPENETRATION

① TEE

DEVELOPMENT

② CORNER

DEVELOPMENT

DEVELOPMENT

241

242

ISOMETRIC VIEW

INTERPENETRATION OF CYLINDERS OF EQUAL DIAMETERS

116

243. Interpenetration of Cylinders Unequal diameters. Right angled. Draw the outline plan and elevation. Draw the generators of the smaller cylinder, obtain their points of intersection with the larger cylinder in the plan. Project these points from the plan to the elevation. The line of interpenetration is obtained by the intersection points of the generators, through which a fair curve should be drawn. The developments of the cylinders should be drawn in line.

244. Two Unequal Diameter cylinders interpenetrating at an angle of 60°.

Draw the outline of plan and elevation. Draw the generators of the smaller cylinder and obtain their points of intersection with the larger cylinder in plan. Project these points to the elevation to intersect the numbered generators there. The curve of interpenetration can be drawn through the points of intersection. The developments can now be made in line as before.

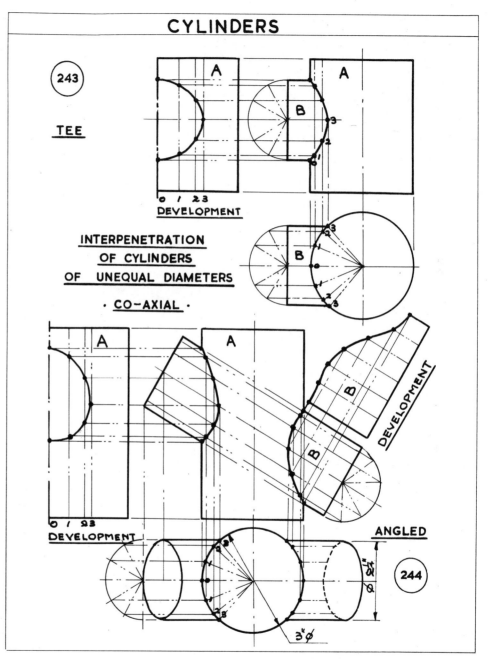

117

245. Cylinders of Unequal Diameter, Offset Draw the plan and elevation. Draw the generators of the smaller cylinder, obtain the points in the plan. Project these points into the elevation to intersect generators which give points on the curves of interpenetration. Note that the front and rear curves are not the same. The developments may be drawn in line with the portion of the cylinder being developed.

The isometric pictorial view shows a horizontal section which gives a circle for the vertical cylinder and a rectangle for the horizontal cylinder. It should be seen that if a number of sections are visualised with intersections of the two shapes at the varying points the line of interpenetration becomes easier to understand. The line of the generators in the elevation show such sections.

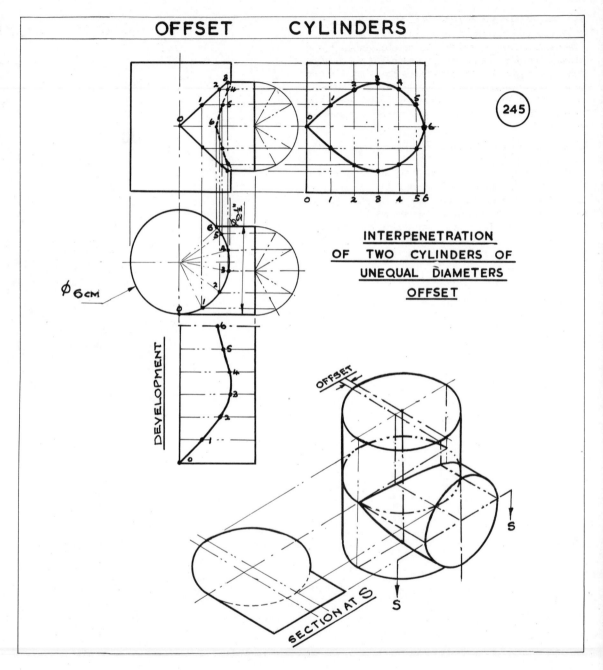

OFFSET CYLINDERS

Ø 6CM

DEVELOPMENT

INTERPENETRATION
OF TWO CYLINDERS OF
UNEQUAL DIAMETERS
OFFSET

OFFSET

SECTION AT S

246. Interpretation of Two Cylinders Unequal diameters. Offset and angled.

Draw the plan and elevation. Draw the generators of the smaller cylinder in the plan and obtain their points of intersection with the circumference of the larger vertical cylinder. Draw the generators of the smaller cylinder in the elevation. Obtain the points of intersection of the generators and thus the line of interpenetration. Note that an additional generator must be drawn for points 1 and 11.

The developments may be drawn in line with the cylinder being developed.

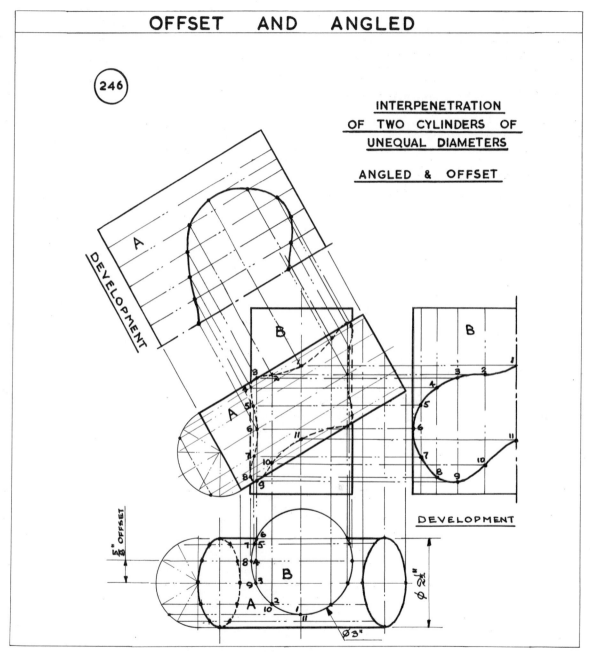

OFFSET AND ANGLED

INTERPENETRATION
OF TWO CYLINDERS OF
UNEQUAL DIAMETERS

ANGLED & OFFSET

247. Two Projections of a Cone and a Pyramid are shown interpenetrated by a cylinder (not shown).

The projections and development are shown. The generators of the cone are drawn from the plan and their points of intersection with the cylinder line projected back to the plan to give the line of interpenetration. The sector of the circle for the development, arc as the twelve divisions of the base circle is drawn, and the points on the generators projected to the line of true length before being swung to the corresponding line in the development.

248. The Pyramid The projections are similar to that of the cone, generators being drawn through convenient points in the cylinder curve in the elevation, and projected to the plan. This enables the line of interpenetration to be drawn in plan. The true length of OC is found by rabatment, this is a generator of the 'enclosing cone' and enables the development to proceed. The base true lengths A B and B C are stepped off along the development curve, and generators drawn in. Points 1 to 7 must be projected to the line of true length before being brought to the corresponding generator in the development.

249. Cylinders and Cones Three cases of interpenetration of these solids are shown. Generators of the cylinders are drawn first, sectional circles of the cone are drawn in the plan, points on intersection obtained, and projected back to the elevation. The line of interpenetration can be drawn through these intersections. Horizontal sections of the solids show the intersection of a circle for the cone and a rectangle for the cylinder.

Refer to the pictorial isometric No. 176.

The developments of the cone are made radially from the generator, the development of the cylinder is best made in line.

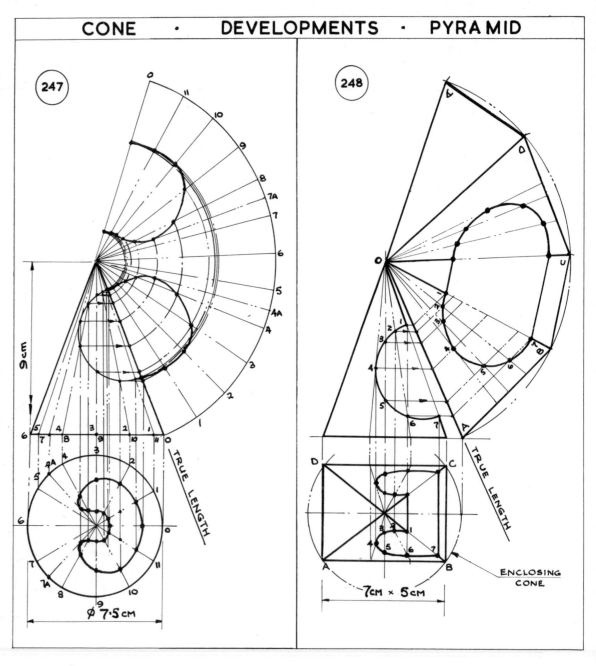

CONE · DEVELOPMENTS · PYRAMID

120

CYLINDER AND CONE

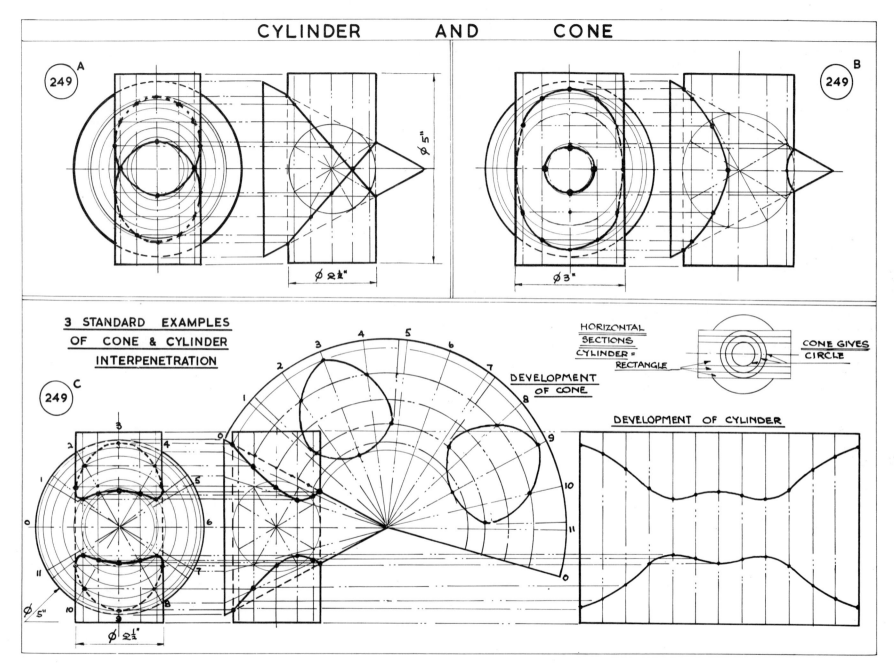

A
(249)

B
(249)

∅ 5"

∅ 2¼"

∅ 3"

3 STANDARD EXAMPLES
OF CONE & CYLINDER
INTERPENETRATION

C
(249)

HORIZONTAL
SECTIONS
CYLINDER =
RECTANGLE

CONE GIVES
CIRCLE

DEVELOPMENT
OF CONE

DEVELOPMENT OF CYLINDER

∅ 5"

∅ 2¼"

250. Cylinder and Cone The interpenetration of a right cone and a vertical cylinder is shown. Draw the traces of four horizontal sections in the elevation, each section line will give a circle for the cone and a circle for the cylinder. The intersection of the circles shown in plan when projected back to the elevation section line will give four points through which the line of intersection may be drawn. Draw First Angle.

251. Cylinder and Sphere Horizontal sections give a constant circle for the cylinder, and varying diameter circles for the sphere. Draw Third Angle.

252. Sphere and Prism A sphere and vertical hexagonal prism are shown. Simple projection gives the intersection of the sphere and the edges of the prism, but a midpoint of the prism must be projected as shown to give midpoint intersection through which the curve may be drawn. Draw Third Angle.

253. Cone and Sphere Horizontal sections give varying circles for both cone and sphere. The projection of their intersections from the plan to the elevation give the line of interpenetration. Draw First Angle.

254. Cylinder and Sphere Horizontal sections give a constant circle for the cylinder and varying circles for the sphere. Projections from the plan will give the interpenetration curve points. Draw First Angle.

255. Two Cones If two cones interpenetrate so that the focal sphere contained touches the four generators, the lines of intersection will be straight. Draw Third Angle.

256. Prism and Pyramid Simple projections give straight line intersections. Draw First Angle.

257. Spherical Ended Cylinder cut by two inclined planes. Horizontal sections give circles in the plan, simple intersection points projected from the elevation give the line of intersection in the plan. Draw Third Angle.

258. Palmate Section If a turned rod is cut by a plane as shown in the diagrams, the curves of intersection are obtained by projection from horizontal sections as shown. Draw First Angle.

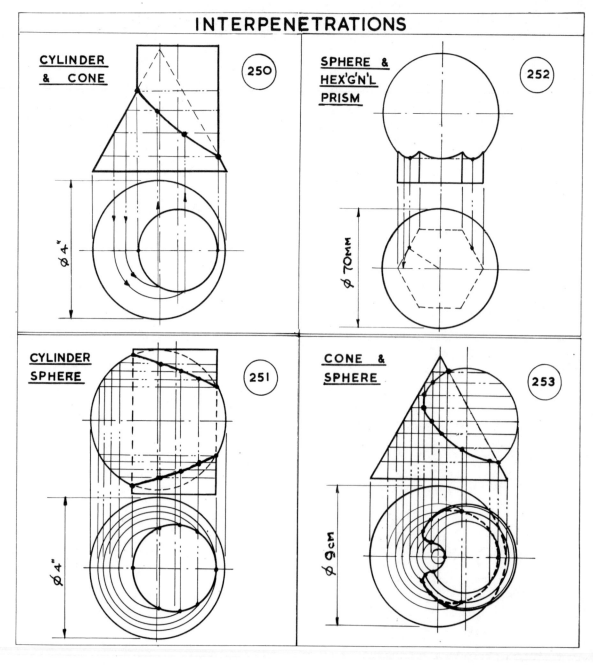

INTERPENETRATIONS

CYLINDER & CONE — 250

SPHERE & HEX'G'N'L PRISM — 252

CYLINDER SPHERE — 251

CONE & SPHERE — 253

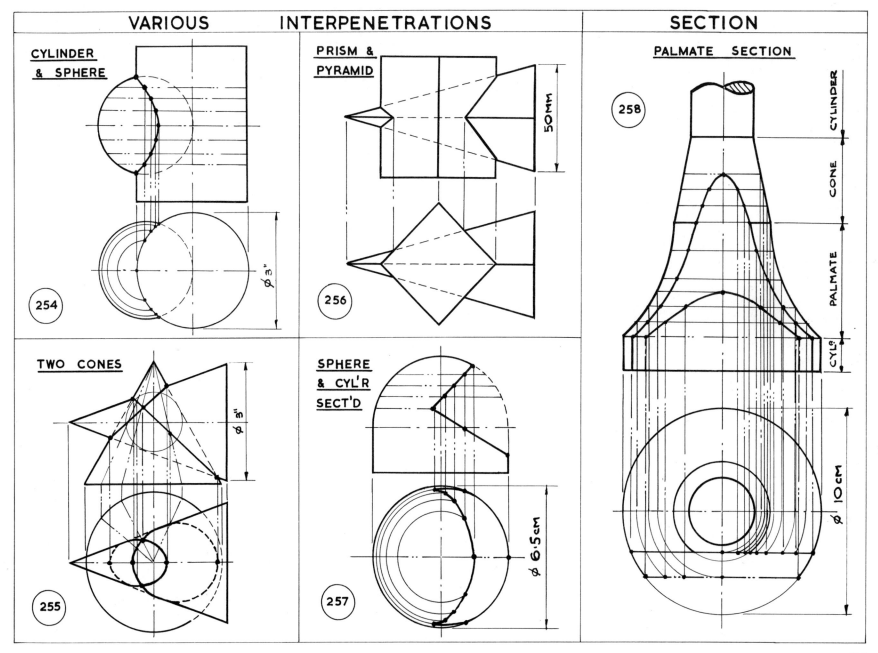

VARIOUS INTERPENETRATIONS | SECTION

CYLINDER & SPHERE

254

PRISM & PYRAMID

50MM

256

PALMATE SECTION

258

CYLINDER

CONE

PALMATE

CYL.

Ø 10 cm

Ø 3"

TWO CONES

Ø 3"

255

SPHERE & CYL'R SECT'D

257

Ø 6.5 cm

123

259. Interpenetrating Spheres The section is a circle, and the projected line of intersection in the given example is a straight line.

260. Touching Spheres Problems of this nature are based on common tangency between adjacent spheres. The distance from one centre to the centre of the touching circle is the sum of the radii. The tangent is at right angles to the line joining the centres. In the example shown of five equal spheres arranged four and one, the four are drawn first touching in plan, then projected to the elevation and the fifth sphere drawn tangentially.

261. Sphere Pierced by Pin True positions on the sphere are found by auxiliary sections in this case on the line of the pin obtained by joining A and B in plan. The section Z/Z enables true heights H_1, H_2 and H_3 to be found and used in the elevation where projectors from the plan allow the projections of the pin to be completed.

262. Cylinder and Touching Spheres Given a cylinder and two spheres draw them in mutual contact. Draw the sphere 1 and cylinder in contact in both plan and elevation. Draw construction circles of the smaller sphere in contact with the cylinder and larger sphere, and swing the centres of the smaller sphere to intersect as arcs at C in plan. The points of contact are then obtained by projection P, P_1, P_2.

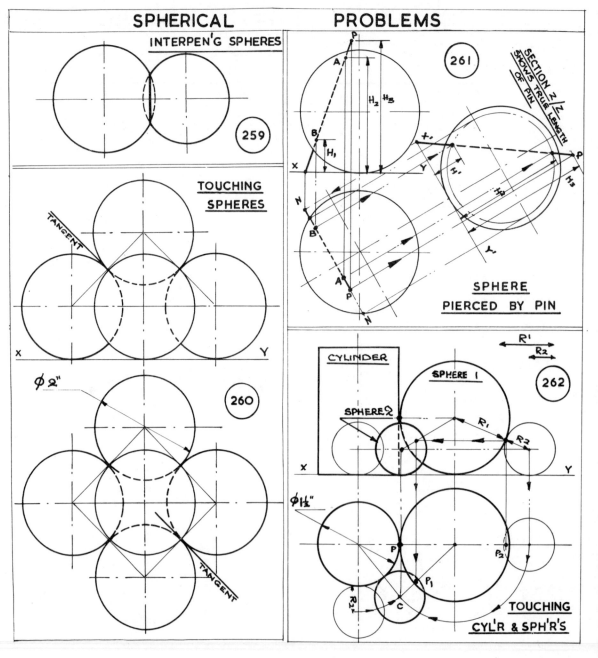

SPHERICAL PROBLEMS

INTERPEN'G SPHERES

TOUCHING SPHERES

SPHERE PIERCED BY PIN

TOUCHING CYL'R & SPH'R'S

263. Spheres in Contact with a Pyramid and each other. The large circle representing the sphere is drawn tangential to the horizontal plane and the side of the pyramid. Point of contact can then be projected to the plan. The smaller circle is then drawn tangential to the H.P. and the pyramid. Before this can be projected to the plan, the two circles must be drawn as in the right of the diagram tangential to each other and to the H.P. obtaining the projected length of R_1 and R_2. This distance is used in the plan to find the centre of the smaller circle. It must be seen that the points of contact are only shown when a mid section of both of the touching spheres is drawn, and often an auxiliary elevation or plan has to be projected first before these circular sections can be drawn.

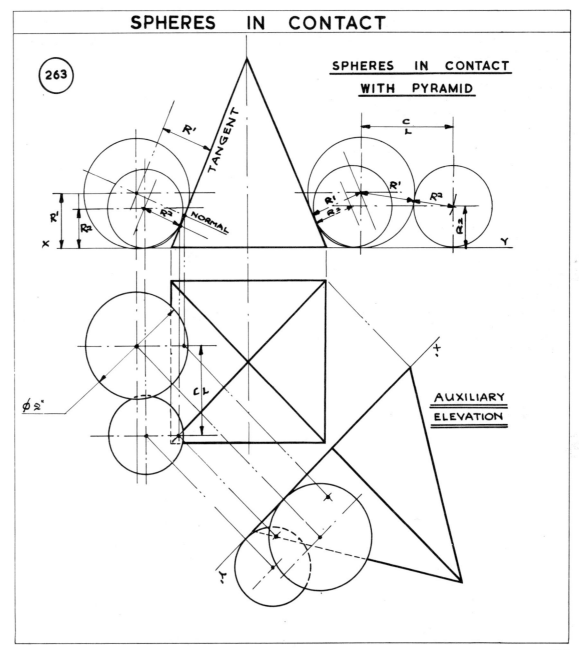

SPHERES IN CONTACT

SPHERES IN CONTACT
WITH PYRAMID

AUXILIARY
ELEVATION

263A. Interpenetration of Cone and Cylinder

A. Given the cone and cylinder sizes, and the angle of interpenetration, find the line of intersection and draw the developments.

Draw the cone elevation. Draw the circle shown equal in diameter to that of the cylinder, and tangential to the slope lines of the cone. The circle represents the 'enveloped' sphere which is common to both cone and cylinder. The generators of both cone and cylinder are tangential to this sphere, and it should be seen that the cylinder can be rotated to any position around the sphere. The centre line of the cone and cylinder can be drawn at the given angle. Draw the outer generators of the cylinder, these will cut the cone. Join these points to give the intersection line which will, in these cases, be a straight line.

The developments can be drawn by methods shown on previous pages.

B. **Given Cone and Line of Intersection**, find the angle and diameter of the cylinder to intersect the given cone. Draw the given cone and line, the given line cutting the cone will yield an ellipse (see 52) which will also be a section of the desired cylinder.

C. Bisect the line of intersection to give C. Cut the cone horizontally through this point, draw the semi-circle on this line. Drop a perpendicular from C to cut the semi-circle, and draw the arc, centre C and radius CB as the perpendicular.

Draw the generators of the cylinder tangential to this arc and to meet the line of intersection as shown.

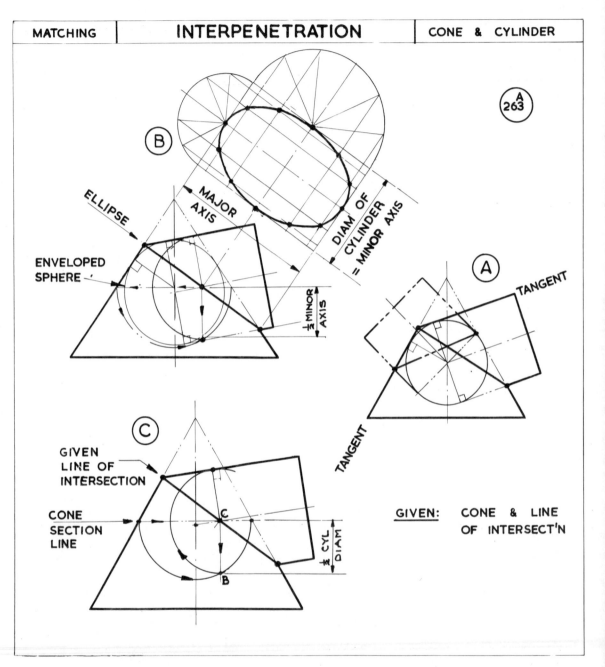

GIVEN: CONE & LINE OF INTERSECT'N

126

FURTHER INTERPENETRATIONS

Angled Prism and Pyramid

Angled Cylinder and Cone

Co-axial and Offset

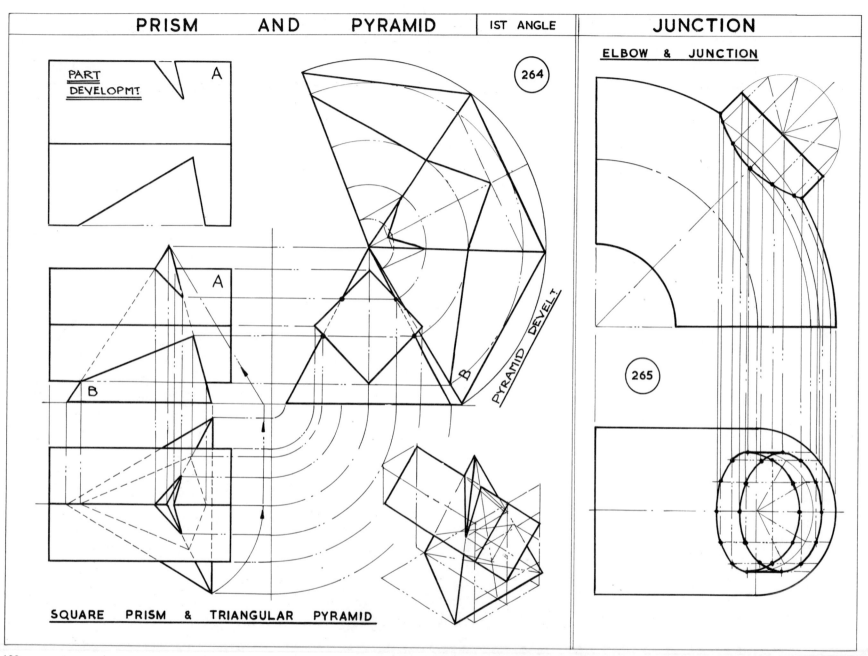

ELBOW & JUNCTION

PART DEVELOPMT.

A

264

B

PYRAMID DEVELT.

SQUARE PRISM & TRIANGULAR PYRAMID

265

264. Square Prism and Triangular Pyramid, interpenetration. The projections and developments are shown. Draw the plan and project the elevation. Points of intersection obtained from the elevations can be projected back to the plan. The 'enclosing cone' must be projected from the plan before the development of the pyramid can proceed.

265. Elbow and Cylinder Junction Projections of a cylinder and a quadrant section of pipe interpenetration are shown. Draw the twelve generators of the cylinder and obtain points of intersection with corresponding horizontal section generators of the quadrant pipe.

266. Hexagonal Pyramid and Angled Triangular Prism. Draw the plan and elevation; all the points of interpenetration can be obtained from the projections, except one at the lower right hand. A generator of the pyramid in plan projected into the elevation to meet the prism line at *a* completes the lines of interpenetration. Points of intersection on the pyramid must be projected horizontally to the true length line which in this case is the outer generator, before being swung to the development. The part development of the prism has been projected in line from the elevation; true lengths at *b* and *c* are found from the true end shape of the prism by a part auxiliary view.

HEXAGONAL PYRAMID & ANGLED TRIANGULAR PRISM   (266)

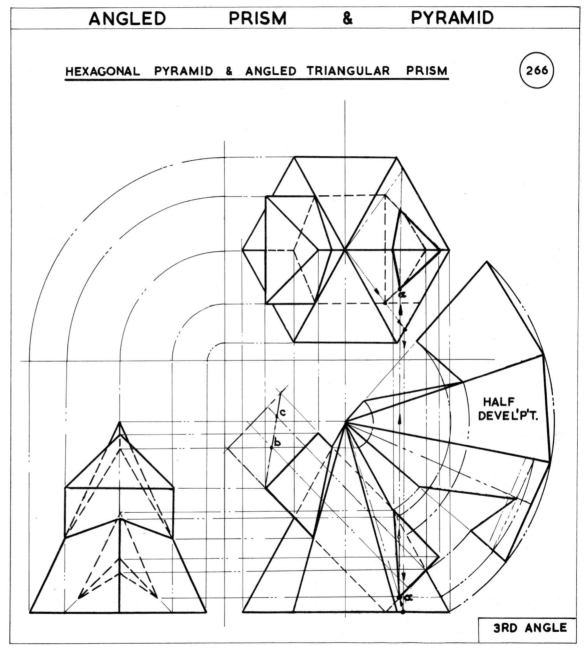

HALF DEVEL'P'T.

3RD ANGLE

267. Cone and Inclined Cylinder Draw the plan and elevation, project the inclined cylinder to the H.P. to obtain the ellipse (Auxiliary plan). Project the apex of the cone to the H.P., and draw several trace lines to cut both circle and ellipse in the H.P. Project these points into the elevation as generators of cone and cylinder. Their intersection in the elevation gives the line of interpenetration, and the curve is drawn through them. The points should be projected back to the plan, and the line of interpenetration drawn. The developments are projected in the usual method, in line for the cylinder and sector for the cone.

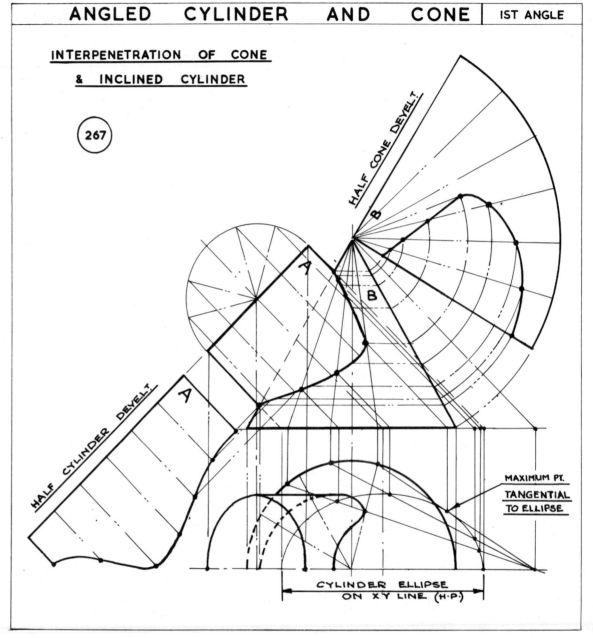

INTERPENETRATION OF CONE & INCLINED CYLINDER

267

HALF CONE DEVELT.

HALF CYLINDER DEVELT.

A

B

A

B

MAXIMUM PT. TANGENTIAL TO ELLIPSE

CYLINDER ELLIPSE ON XY LINE (H·P·)

268. Interpenetration of Cone and Inclined Prism. First angle projections of the interpenetration are shown. Draw the plan and elevation. Project an auxiliary plan at right angles to the axis of the prism. A true section of the prism will be shown. Widths below *xy* and x^1y^1 lines are the same in both original and auxiliary plans. Draw the generators of the cone in the plan, project to the elevation and thence to the ellipse in the auxiliary plan. Number the points of intersection of the prism section with the generators in the auxiliary plan. Project these points to the elevation and to their respective generator. Join these points in the elevation to give the curve of interpenetration. Project the points to the original plan to the respective generator. Join these points to give the line of interpenetration in the plan.

Draw the development of the cone. Draw in the generators. Mark off the true length of the point of intersection on each generator, draw a fair curve through the points to complete the development of the cone.

Project the development of the prism at right angles to the axis of the prism.

Make the developments in thick paper and assemble as a model to prove the method.

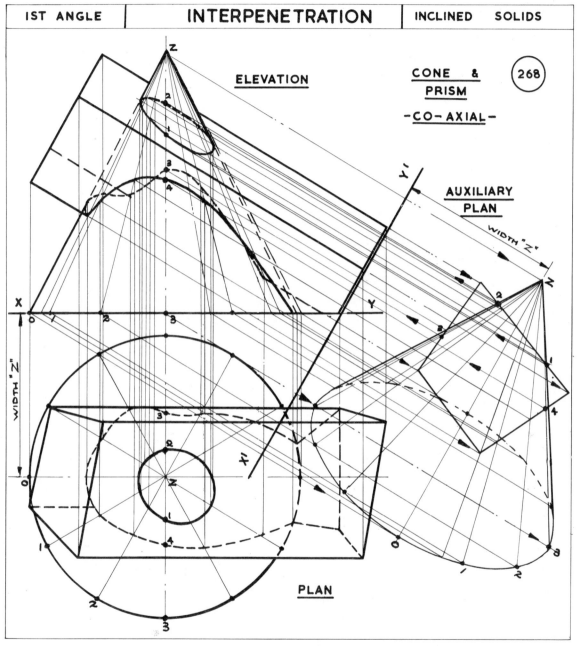

ELEVATION

CONE & PRISM

268

-CO-AXIAL-

AUXILIARY PLAN

WIDTH "Z"

WIDTH "Z"

PLAN

269. Interpenetration Oblique Octagonal Pyramid and an Inclined Cylinder-offset.

Third Angle projections of the interpenetration are given.

The problem may be set showing the incomplete plan and elevation, or by description of the two solids and the detail of the offset axes.

Draw the incomplete plan and elevation in simple projection, label the xy line.

Project an Auxiliary Plan as shown, using a new x^1y^1 line which is normal (at right angles to) the axis of the inclined cylinder. Widths remain the same in the new plan, i.e. points 0, 1, 2, 3, 4, 5, 6, 7 will be the same distance from x^1y^1 as from xy, in their respective plans.

The auxiliary plan will show points of intersection of the cylinder with the faces of the pyramid.

These points may now be projected back to the respective generator in the elevation, giving points through which the line of interpenetration may be drawn.

The line of interpenetration in the original plan may be obtained by projecting the points from the elevation to this view and drawing a curve joining the points.

Note that the points will be a similar distance from their respective xy line and this gives an additional check on accuracy in projection.

Developments of the cylinder and pyramid can now be made to prove the method.

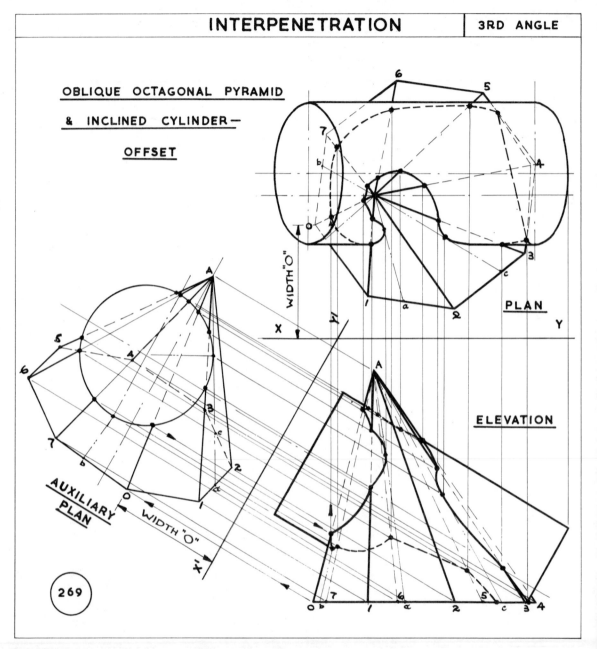

OBLIQUE OCTAGONAL PYRAMID
& INCLINED CYLINDER—
OFFSET

PLAN

ELEVATION

WIDTH "O"

AUXILIARY PLAN

WIDTH "O"

269

132

270. Interpenetration   Oblique Cone and Inclined Cylinder, offset.

First Angle projections of the interpenetration are given. The problem may be set by incomplete views or by a written description.

Draw the views as fully as possible from the given data. Project an Auxiliary Plan from the elevation on x^1y^1, normal (at right angles to) the axis of the cylinder. Widths remain the same as in the original plan, note width 'Z'.

Draw the generators of the cone, project these to the Auxiliary Plan. Where these cut the circle which represents the section of the cylinder, points 1, 2, 3, 4, 5, 6, 7, 8, 9, 10, 11, 12 and a, b, c, are found. These are projected back to their respective generators in the original elevation. The fair curve drawn through the points gives the line of intersection.

The positions may now be projected to the original plan and the line of intersection shown in this view.

Check that the points in both plans are the same 'widths' from the respective xy lines.

Draw developments of the cylinder and cone on strong paper or card with construction flaps and erect the model.

Reference to the development of the oblique cone and pyramid by the triangulation method shown in the next section should be made.

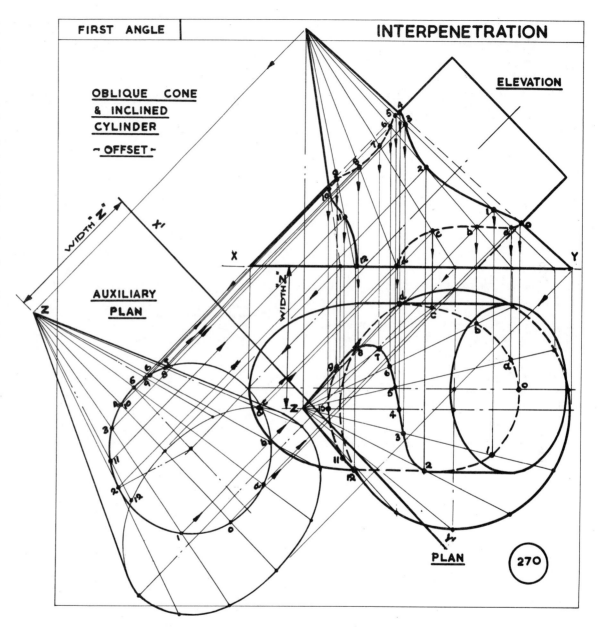

FIRST ANGLE

INTERPENETRATION

OBLIQUE CONE
& INCLINED
CYLINDER

~ OFFSET ~

ELEVATION

AUXILIARY
PLAN

PLAN

270

Exercises

271. Simple Developments

A. Scale up the section of irregular hexagonal tube shown. Project the development.

B. Draw the development of the section shown.

C. A cut portion of a pentagonal tube is given. Draw its development.

D. Develop the sides of the square hopper shown.

E. Develop one side of the canister shown.

F. Develop one side and one end of the box shown.

G. Draw the developments of the throat piece and one cylindrical portion shown in the diagram.

H. Draw the projections of the two interpenetrating cylinders shown whose axes are offset by $\frac{1}{4}''$. Draw the developments of the two cylinders.

I. Three cylinders of equal diameter meet as in the diagram. Draw the projections showing the line of interpenetration. Draw also the developments of the cylinders.

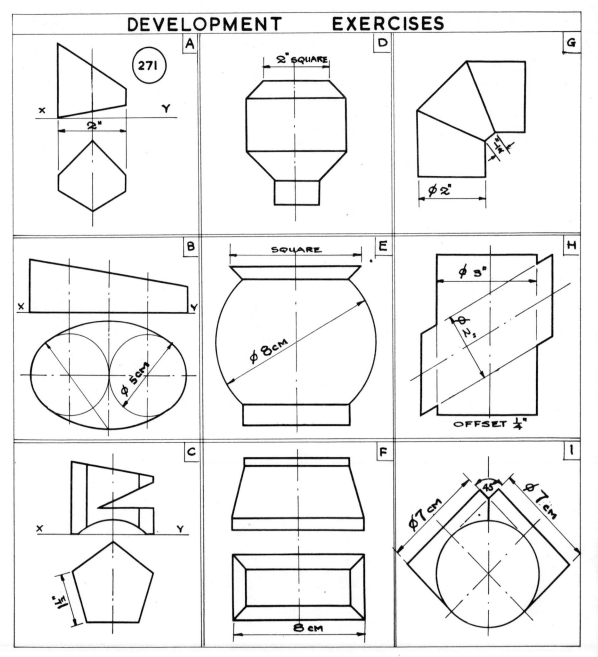

DEVELOPMENT EXERCISES

134

Exercises

272. Interpenetrations and Developments

A. Draw the developments of the two cylinders shown in the elbow.

B. Draw the developments of the two pieces of tubing shown in the elbow.

C. Draw the projections of the two offset unequal diameter cylinders which meet at right angles. Draw the developments of the two cylinders.

D. A cylinder and cone interpenetrate at 90°, co-axial. Draw the projections and show the line of interpenetration. Draw the developments.

E. A right cone and cylinder interpenetrate as shown, co-axial. Draw the full projections, and developments.

F. A cone and square prism interpenetrate at right angles as shown. Draw the full projections, and draw the developments.

G. A square pyramid and a sphere interpenetrate as shown in the diagram. Draw the projections, Draw the development of the pyramid only.

H. A sphere and cylinder interpenetrate as shown. Draw the projections; project the development of the cylinder.

I. A hemisphere and two cylinders interpenetrate co-axial as shown in the diagram. Draw the plan and completed elevation. Draw the developments of the two cylinders.

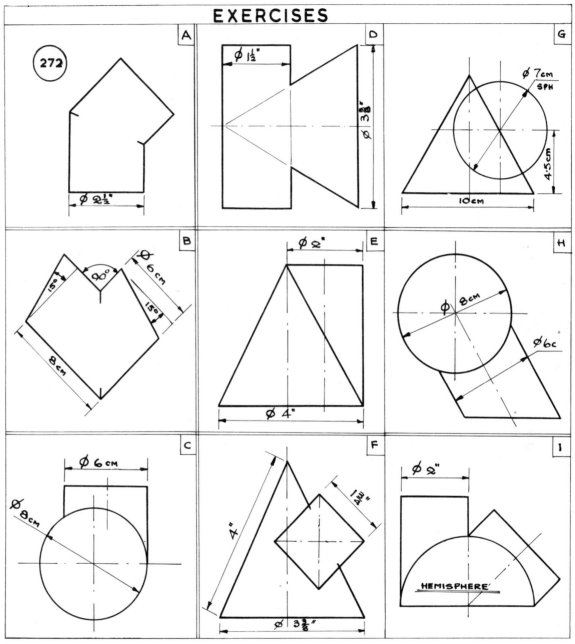

EXERCISES

135

Exercises

272A. Interpenetrations

A. A cone and cylinder interpenetrate as shown. Draw the full elevation and plane. Draw the developments, in First Angle.

B. A cylinder and triangular prism interpenetrate, angled and offset as shown. Draw the completed plan and elevation. Draw the developments, in Third Angle.

C. A cylinder and two cones interpenetrate as shown. Draw projections which show the line of intersection of surfaces, and draw the developments, in First Angle.

D. A triangular pyramid and cylinder interpenetrate as shown in the diagrams. Complete the line of interpenetration, and draw the developments, in Third Angle.

E. From the figure shown, draw the projections and developments of the four parts, in First Angle.

F. An octagonal pyramid and a cylinder interpenetrate as in the diagram. Complete the projections, and draw the developments, in Third Angle.

G. Two cones interpenetrate as shown. Draw the full projections and developments, in First Angle.

H. A cone of 4″ diam. base circle and 4″ height is pierced by a triangular prism positioned as in the given plan. Complete the projections, and draw the developments, in Third Angle.

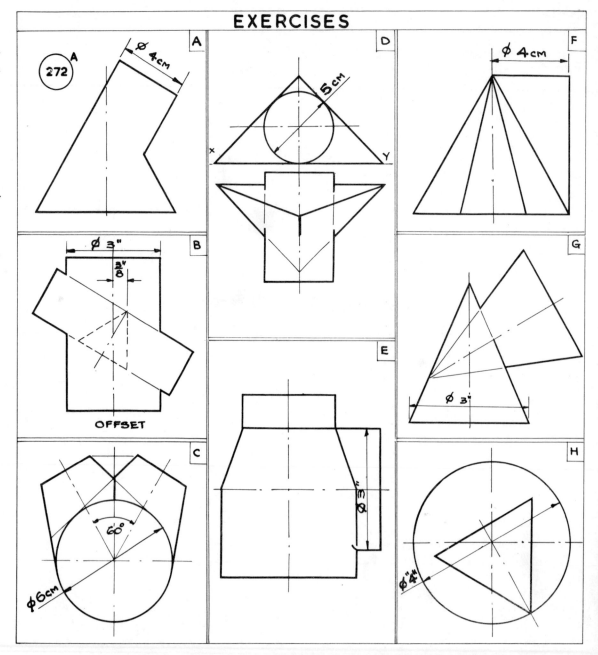

DEVELOPMENT BY TRIANGULATION

Oblique Pyramid

Oblique Cone

Transition Pieces

Junction Piece

273. Oblique Pyramid The true lengths of AO, BO, CO and DO are obtained by rabatment from the plan. Arcs are swung from O in the plan to a line parallel to the *xy* line, from A, B, C and D, and rabatted to the *xy* line. A¹, B¹, C¹, D¹, are then joined to O¹ giving the true lengths of the corners of the pyramid.

The true lengths should be seen to be the hypotenuse of a right-angled triangle which has been rabatted to the vertical plane so that its true shape may be seen. This system is shown in the chapter on Lines on the Inclined Plane.

The true lengths of AB, BC, CD and DA are given directly from the plan. The development of the pyramid is now drawn by constructing, by the arc method, the four triangles shown, the first, for example, has sides using true lengths, O¹C¹, O¹D¹, C¹D¹.

Where the pyramid is truncated, the true lengths of the corners may be obtained by horizontal projectors from the elevation cutting the true length lines as shown, and transferring these points to the development.

274. Transition Piece Square to square, angled. From the plan and elevation given it will be seen that the faces are triangles. The bases of the triangles are rabatted as before to the elevation, where heights remain the same, and the new hypotenuse gives the true length of the line required. Lengths of the squares 1, 2, 3, 4, and A, B, C, D, are truly given in the plan, and the development may be drawn as shown.

Notice that in this example, there are two sets each of four similar triangles. The examples should be drawn to a convenient size, say twice the printed size, cut out and pasted together (make pasting flap at 3A), as proof of the method.

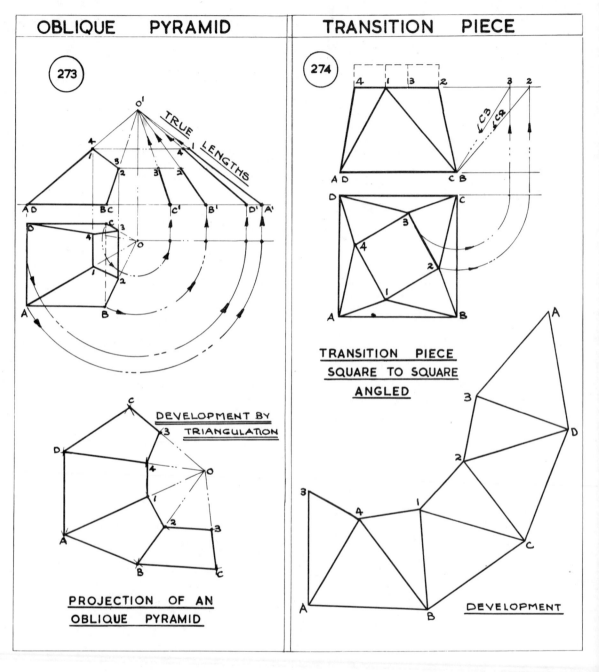

OBLIQUE PYRAMID

273

TRUE LENGTHS

DEVELOPMENT BY TRIANGULATION

PROJECTION OF AN OBLIQUE PYRAMID

TRANSITION PIECE

274

TRANSITION PIECE
SQUARE TO SQUARE
ANGLED

DEVELOPMENT

275. Transition Piece Rectangle to circle. Pipes and ducts for conducting air, gases, powdered solids, etc., often require two dissimilar sections to be smoothly joined usually by a transition piece such as the one shown. A pattern of the development has to be made before cutting, folding and joining the sheet metal to form the transformer piece.

The method used is that shown in the previous example, triangulation, the circle being divided into twelve parts, the points being joined to the nearest corners of the rectangle to give connecting triangles. The true shape of the triangles is then obtained by rabatment, and the final development drawn by constructing the triangles in their adjacent order.

The plan and elevation of the piece is given in Third Angle projection. A B, B C, C D, D A are true lengths as shown in the plan. 0 1, 1 2, 2 4, etc. are chordal. B O, B 1, B 2, etc. are rabatted as shown, and a new hypotenuse drawn to the height line from the elevation gives the true length. The development is now drawn by constructing the triangles in the correct order, using true lengths.

The piece should be drawn to a convenient size, take A B as 4″, cut out and pasted together as an example.

Chordal distances on the circle are usually allowed to be taken as the third side of the triangle, but accurate work would demand that the circle be taken as D = 3·1416, and divided by 12 to give the true length of the small arc.

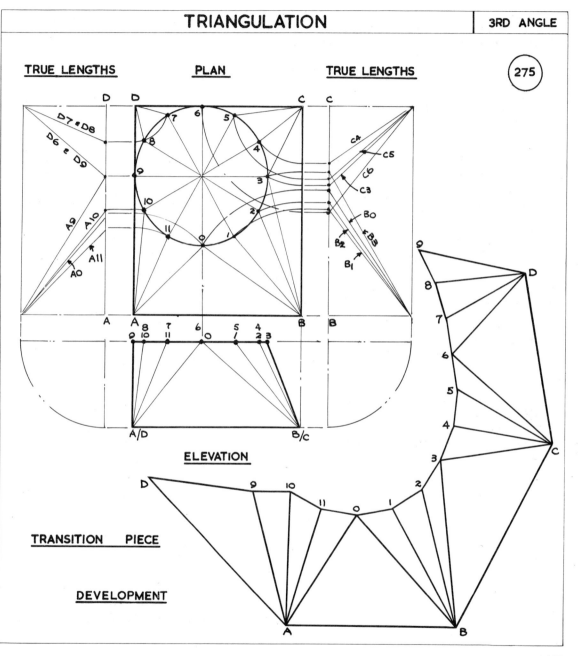

TRUE LENGTHS   PLAN   TRUE LENGTHS

ELEVATION

TRANSITION PIECE

DEVELOPMENT

276. Transition Piece Square to circle. Draw the plan and elevation. Divide the circle into twelve parts, join as in the diagram to A B C D. Obtain true length of C_6 and C_7 by rabatment to the elevation. Draw the development by constructing the triangles. Lengths $0\,1$, $1\,2$, are chordal.

277. Transition Piece Shaped hood. Draw the plan and elevation. Divide the circle into twelve parts. Join to the shown points A B C D E F G, to give triangles. Draw the rabatted triangles in the elevation to obtain true lengths of A_1, B_1, B_2, etc. In the diagram, in this instance, the bases of the rabatted triangles have been stepped off by dividers to the elevation instead of arcs from the plan. The hypotenuse of the triangle is the required true length to be used in drawing the triangles in the development. Begin the development by drawing O A as a centre line since the shape is symmetrical about this line.

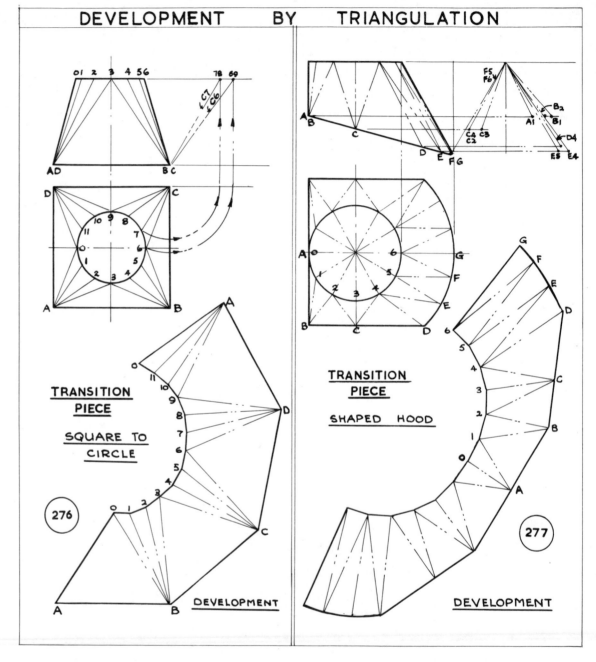

DEVELOPMENT   BY   TRIANGULATION

TRANSITION PIECE

SQUARE TO CIRCLE

(276)

DEVELOPMENT

TRANSITION PIECE

SHAPED HOOD

(277)

DEVELOPMENT

140

278. Oblique Cone Notice the circular plan of an oblique cone.

Divide the base circle into twelve parts. Draw generators. Drop a perpendicular from the cone apex A to cut the *xy* line in A^1.

Rabat points 0 to 6 to the *xy* line to give points 0^1 to 6^1. Join these to A. These are the lines of true length of the generators shown.

The full development of the oblique cone may now be drawn by constructing triangles using true lengths obtained above, and the chordal distances from the base circle. (True length of the arc is given by D = 3·1416/12 for more accurate work if required.)

Three section lines of the cone are shown. True lengths from the section lines can be obtained by projecting horizontally from the intersections of the section lines and the respective generator to the correct line of true length. These distances are now transferred to the respective line in the development and joined by a fair curve.

Notice the shape of the development of the oblique cone as compared with that of the right cone, shown on an earlier page.

Draw the example of the oblique cone shown on this page, base circle 3″ diam., height 2$\frac{3}{4}$″, generator A^3 at 45° to the *xy* line. Draw the development when the cone is cut by a plane parallel to the H.P. and 1$\frac{1}{4}$″ above it.

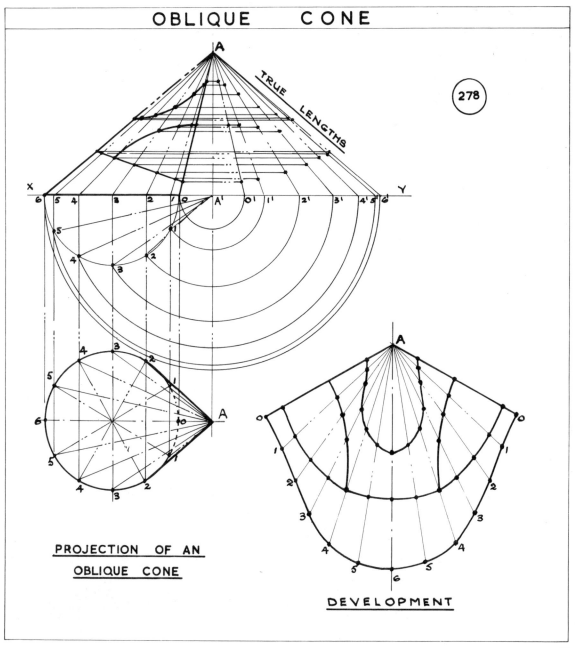

OBLIQUE CONE

278

PROJECTION OF AN
OBLIQUE CONE

DEVELOPMENT

141

279. Junction Piece Three cone. Third Angle projections of a junction piece in which three cones meet at 120°.

Draw the plan and elevation as shown, using a base circle of 3″ diam., cone height 3″, upper circle of cone $1\frac{1}{2}$″ diam.

Draw the generators. Obtain the points and lines of intersection.

Draw the part development shown (a) by simple conical development, using the apex of the cone, (b) by triangulation.

In (a) find the apex of the cone by extending the two extreme generators. Obtain the true length of $A0^1$, and draw the full development of one cone. Draw horizontal projectors from the intersection points in the elevation to the line of true length $A0^1$, and cut the generator lines in the development by arcs from A. Join these points by a fair curve to give the final development.

In (b) set out the diagram of part of a cone as shown; obtain first the true lengths of 00^1, 11^1, 22^1 and 33^1 by triangulation as in the previous examples. Draw in the diagonals of the three panels of the cone shown, and obtain their lengths by triangulation. Check the lengths on the first development.

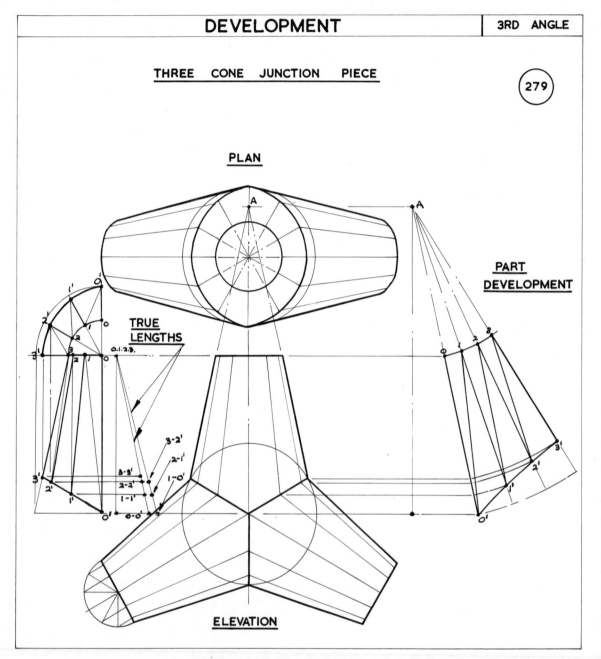

THREE CONE JUNCTION PIECE

279

PLAN

PART DEVELOPMENT

TRUE LENGTHS

ELEVATION

Exercises

280. Development by Triangulation

A. Make the development of the oblique cone shown.

B. Make the development of the oblique octagonal pyramid shown.

C. A transition piece circle to square is shown. Make a development of the surface of the piece. Joint to be on centre line of the front face.

D. A hood is shown; develop the surface by the method of triangulation, joint to be left rear corner seen on the plan.

E. Develop the transition piece shown by triangulation, joint to be on the rear centre line.

F. Make a development of the surface of the truncated oblique cone shown in the diagrams. Joint to be on the shortest generator.

G. A transition piece, rectangle to circle is given. Develop the surface by triangulation. Make the joint on the centre line of the left face.

H. Make a development of the hood shown in the diagrams.

I. Develop the fairing shown in the diagrams. Make the joint in the middle of the front face.

The projections given in the above problems should be scaled up from the one dimension given. The developments could be made finally on card cut-outs to prove the answer.

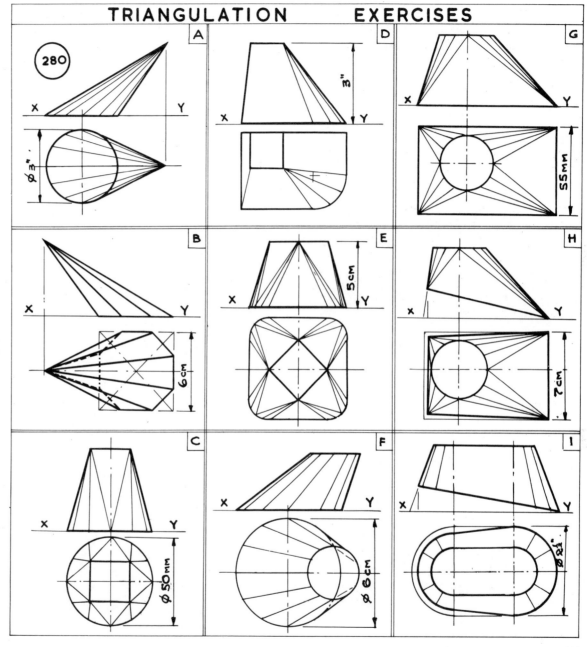

TRIANGULATION EXERCISES

Exercises

280A. Development by Triangulation

J. A transformer or transition piece is shown. Draw the views as shown, draw the generator lines. Obtain the lines of true length. Draw the development.

K. A hood is shown, circular at the base, transforming to a rectangle at the top. Draw the given views, draw in the generators. Obtain the lines of true length, and proceed to construct the development.

L. A transformer piece is shown, base shape encloses three circles, changing to a circular top. Draw the given views, draw the generators. Obtain the lines of true length, construct the development.

M. A shaped hood is shown, circular base, changing to an angled square. Draw the given elevation, project a plan, draw in the generators. Obtain the lines of true length, construct the development.

The developments should be cut out and pasted together to prove the solution.

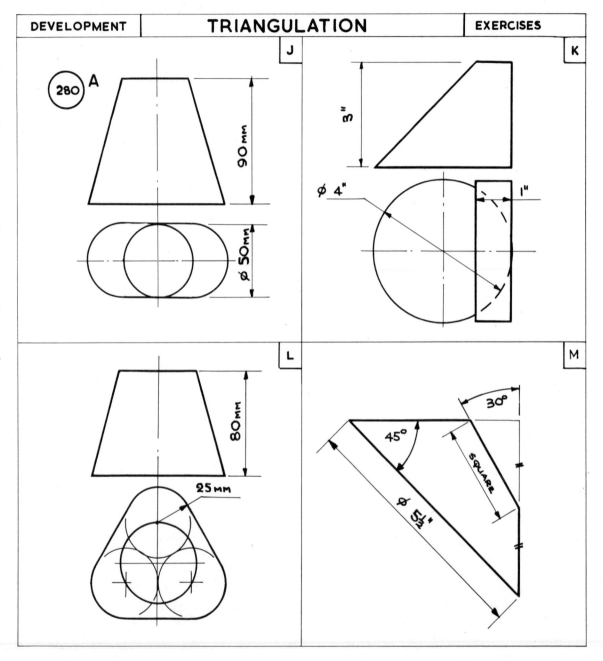

DEVELOPMENT | TRIANGULATION | EXERCISES

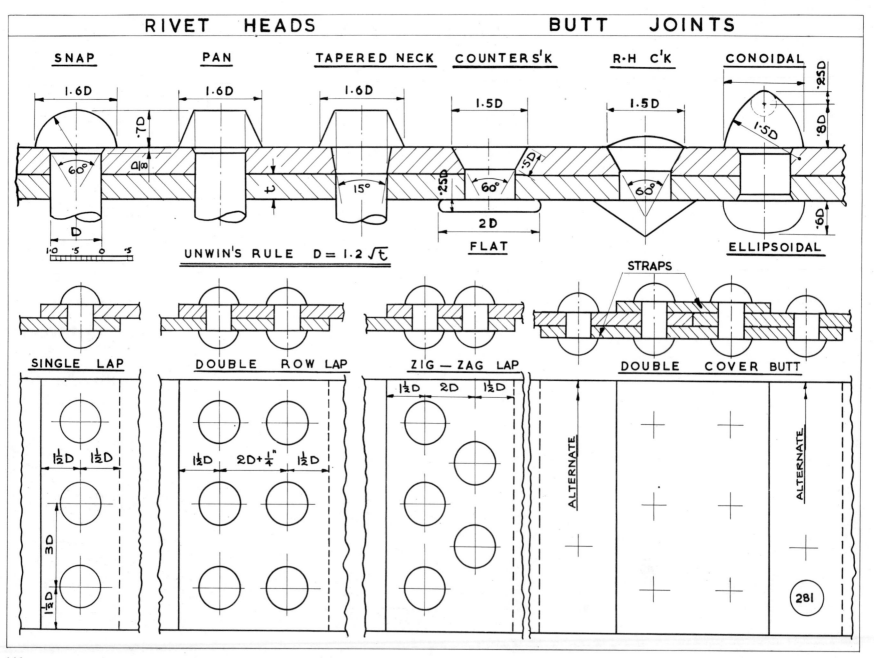

SNAP   PAN   TAPERED NECK   COUNTERS'K   R·H C'K   CONOIDAL

UNWIN'S RULE $D = 1.2\sqrt{t}$

FLAT

ELLIPSOIDAL

STRAPS

SINGLE LAP   DOUBLE ROW LAP   ZIG — ZAG LAP   DOUBLE COVER BUTT

ALTERNATE

281

281. Rivets A rivet is a permanent method of fastening two or more pieces of metal together. The head is pre-formed by stamping, the second head closed by pneumatic hammer or hydraulically whilst red-hot. Small rivets are closed by hammer and set. The shape and detailed form of common rivet heads are shown, snap, pan and countersunk being most used.

Unwin's Rule Given the thickness of the plate to be joined, the diameter of the rivet shank can be found by applying the formula, $D = 1.2\sqrt{t}$.

Rivet Spacing A simple spacing based on the diameter of the rivet is shown for single row lap riveting, for double row chain riveting, for zig-zag riveting, and for a butt joint with double cover straps, the outer row left alternate.

282. Seven riveted joints are shown. Corner joints in tanks may be made by bending the plate or by the use of angle. Slides and box constructions can be built up from standard sheet, strip, angle, channel and tee sections.

283. Joint Failures A rivet may fail by single or double shear stress. A riveted joint may fail by the plate tearing on the line of the rivets, or by marginal tearing, or by elongation of the rivet holes.

Stresses in Rivets These may be calculated by using the following formulae, where d = rivet diam., p = rivet pitch, t = plate thickness, f_t = safe tensile stress for plates, f_s = safe shear stress for rivets, f_c = safe crushing stress for rivets or plates. (Rivets, 25 to 30 tons/in².)

Single shear = $V_s = \pi d^2 f_s/4$.
Double shear = $V_d = \pi d^2 f_s/2$.
Crushing = $V_c = dtf_c$.
Tearing resistance = $(p-d)tf_t$.

d may be found by equating f_s and f_c,

$$d = \frac{4tf_c}{\pi f_s}$$

p may be found by equating tearing resistance and shearing

$$p = \frac{\pi d^2 f_s}{4tf_t} + d \times \text{number of rivets} \times \text{rows}.$$

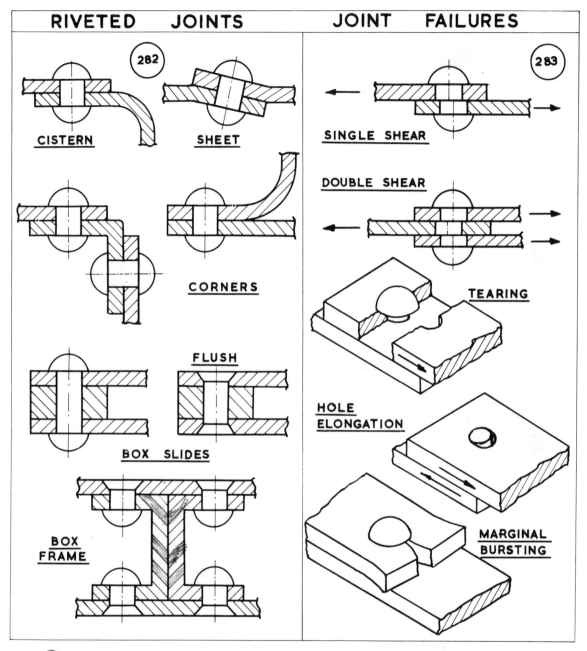

RIVETED JOINTS

CISTERN SHEET

CORNERS

FLUSH

BOX SLIDES

BOX FRAME

JOINT FAILURES

SINGLE SHEAR

DOUBLE SHEAR

TEARING

HOLE ELONGATION

MARGINAL BURSTING

Box Frame

147

284. Riveted Structures The structural engineer uses standard mild steel hot rolled sections for the framework of buildings and roofs. The members are cut to length, drilled or punched for the rivets or bolts and assembled.

The diagram shows a stanchion with its soleplate riveted in position using plate gussets, and angle pieces. The joists are fixed in position with brackets made from angle sections, and the top plate is also bracketed by angle sections.

285. Roof Trusses Designs for these are shown in another chapter, and the frameworks are made in angle or other section, riveted together by means of plate gussets, as shown in the sketches.

Sections of angle, tee, channel and beam are shown, these are made to standard sizes and sections.

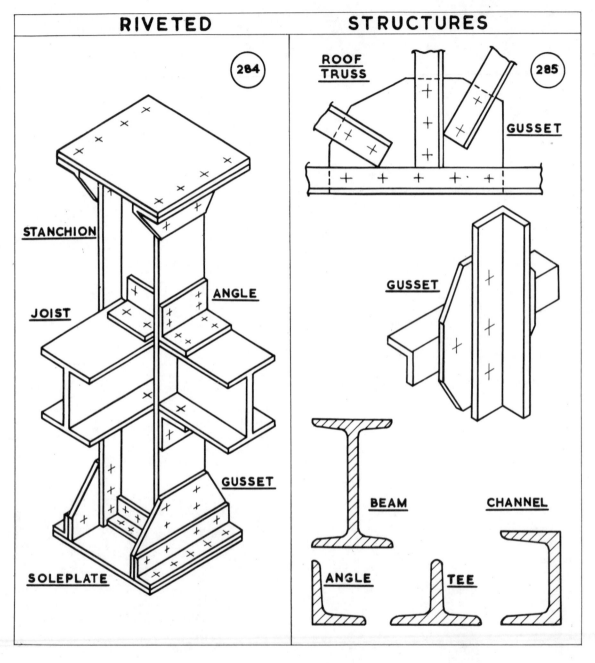

RIVETED STRUCTURES

284

STANCHION

JOIST

ANGLE

GUSSET

SOLEPLATE

285

ROOF TRUSS

GUSSET

GUSSET

BEAM

CHANNEL

ANGLE

TEE

286. Special Rivets for thin plates. Thin plates require rivets which can be closed without distorting the sheet. Aluminium and duralumin, extensively used in the aircraft industry, not easy to solder and weld, are riveted using rivets shown in the diagrams.

Tubular	Short lengths of tube set by a shaped plunger.
Mushroom	Large head, second end closed to the flat shown.
Snap	Solid or part hollow. A small explosive charge detonated by a hot plunger enables the rivet to be set in difficult positions.
Flush	Countersunk rivets giving a flush or level surface. Where the sheet is stamped or 'dimpled' to form the countersinking a stronger joint results.
Mandrel Type	Where the plate is inaccessible from one side, the hollow rivet is expanded by means of a mandrel which is then cut off level. (Avdel type).

The Chobert hollow rivet allows the mandrel to be withdrawn after expanding the rivet, a plug then sealing the hole if required.

Two forms of rivet spacing are shown in addition to those previously given. The first shows alternate spacing in the outer row of rivets, the plate being less weakened. The second shows a three, two, one arrangement for a strap end which also tends to weaken less.

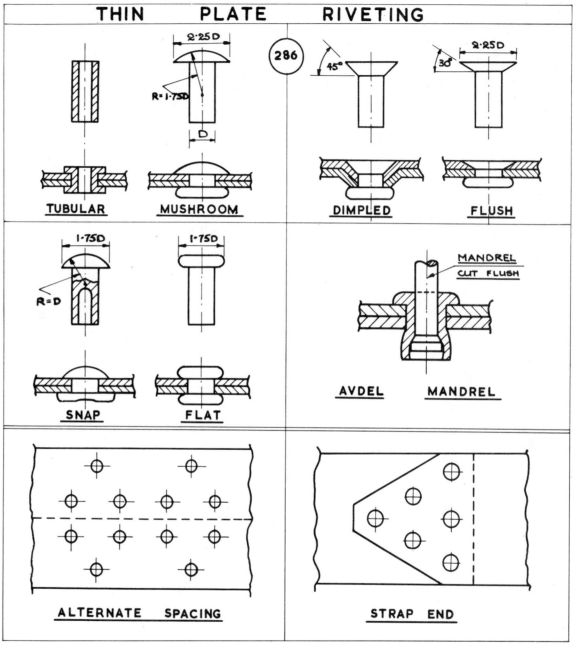

287. Screw Threads Nuts and Bolts. Simplified conventional forms for rapid rendering of external threads of bolts and studs are shown, and a section of an internal thread. First Angle projections of a standard nut and bolt show the conventional representation to accepted proportions based on D = Diameter of bolt. The head may be drawn by drawing the basic hexagon tangent to a chamfer circle $1\frac{1}{4}$D = $\frac{1}{4}$″ (A E B allows the hexagon to be drawn inside a 2D circle), and projecting the elevation of the bolt, making the head $\frac{3}{4}$D in thickness. The nut is $\frac{7}{8}$D in thickness; the diagram shows how the arcs for the facets are obtained.

Conventional details of the nut and bolt in position show it in full and the casting parts in section. The stud, nut and washer (2D + $\frac{1}{8}$″ in diam.) are shown also in full whilst the casting is sectioned and the bolt is rendered similarly.

Note how clearance of the thread is shown in the tapped hole; note how the thread on the bolt and stud allows for tightening.

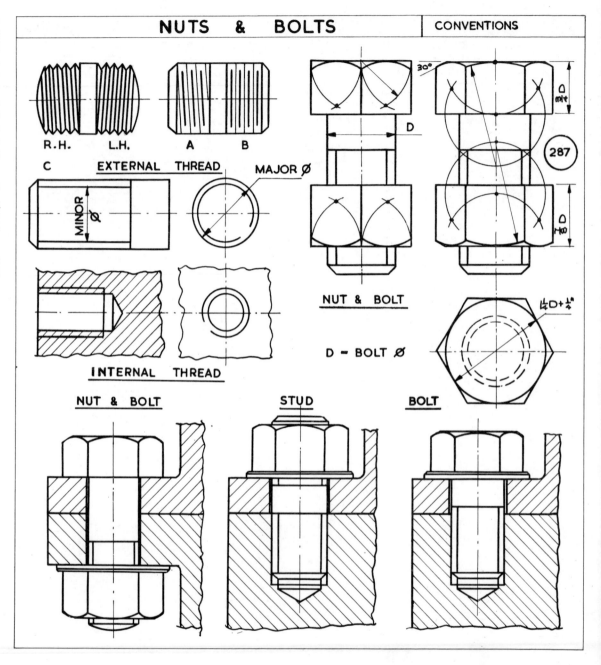

150

288. Screwthread Forms Sectional views of a number of screwthread forms are shown.

The thread follows the path of a helical groove cut or pressure rolled on the cylinder of the bolt or stud, and may be right-handed or left-handed, single or multi-start (see helices). The pitch is the distance measured from crest to crest of two adjacent thread forms. The major diameter is the full diameter over the crests. The minor diameter is the core or root diameter.

British Standard Whitworth Threads (B S W). Standard engineering thread form. Fairly coarse form. Standard Fine is similar but for finer pitch. See tables.

British Association Thread (B A). A millimetre thread used for instrument work. Sizes 0 B A to 25 B A.

American or Sellers Thread Basic American Standard Thread in both coarse and fine forms.

Unified Thread The unified thread form has been accepted by North America and Britain in both coarse and fine form. Three classes of fit are allowed, 1A, 2A, 3A external thread, for ordinary, good class and precision work respectively. 1B, 2B and 3B denote internal threads. An instruction could read: $\frac{3}{4}$ 12 U N F 2A to indicate a $\frac{3}{4}''$ diam., twelve threads per inch, good class Unified external thread.

Metric Thread A 60° angle thread similar to the Sellers thread, sizes in millimetres. See I S O Appendix I.

Special purpose threads, **Square.** Used for threads taking pressure in both clockwise and anti-clockwise direction. Used extensively in machine parts for operating slides.

Acme threads used for lead screws.

Buttress threads. Use where pressure is exerted in one direction.

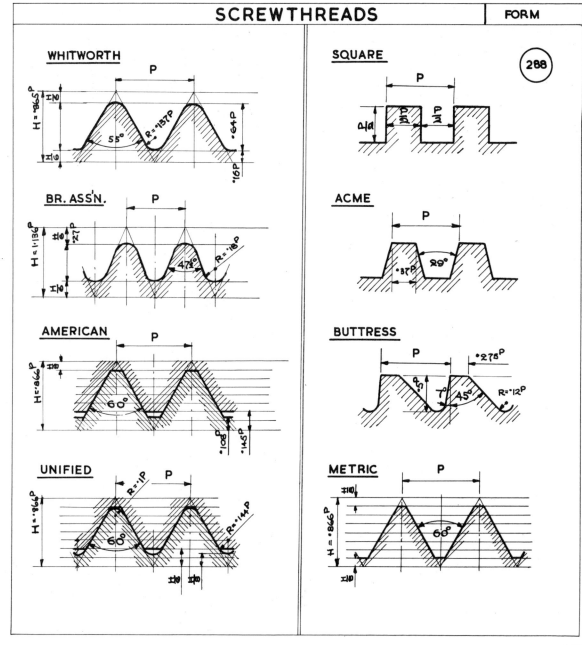

289. Bolts and Screws Various types of bolts and screws with proportions for conventionally rendering in drawings are shown.

Roundhead, cheesehead, fillister, countersunk and instrument screws are slotted for screwdriver turning, and consequently are used in smaller sizes in lighter work.

The square and teehead bolts are used often in slotted slides and faceplates when fixing work to be machined, the head and neck preventing rotation, the bolt then becoming a moveable stud.

The cheesehead bolt may have a short dowel drilled into the shank beneath the head to prevent rotation.

Grub screws may have square heads for tightening which may or may not be sawn off after insertion. Allen grub screws are tightened by means of a hexagonal key fitting into a similar shaped socket in the screw. Headless grub screws are much used for securing pulleys without projection and imbalance.

Clevis bolts are used as pins in forks, entering a plain hole in the second side and screwing into the first side.

Self-tapping screws are frequently used for connecting thin metal sheets. The slotted, case hardened thread cuts its own thread in the softer metal sheet after a drilled or punched hole has been made.

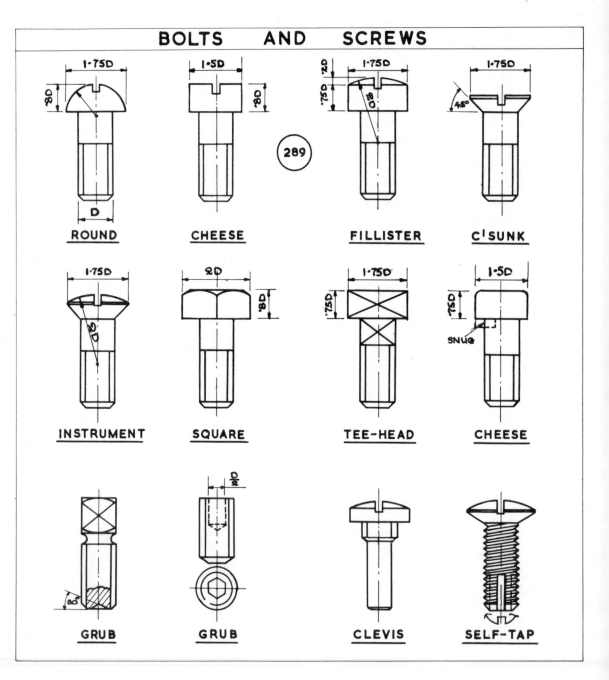

152

290. Nut Locking Devices

Locknut A nut may be secured on a stud against vibration by the use of another nut as shown.

Slotted nut The nut may be secured by using a taper pin or a split cotter pin which passes through a hole in the stud and lies in a slot cut in the nut.

Castle Nut Similar to the slotted nut, the turned reduction allows the folded ends of the split pin to lie without projection.

Ring Nut A set screw is tightened into the turned annulus after the nut has been tightened. A taper pin may also be driven through the end of the bolt as an additional safeguard.

Ring Nut Where a ring nut is used away from the edge of a plate or casting, an additional collar has to be used and located by a dowel pin as shown.

Wile's Nut This invention adopts a slit nut which can be compressed by a set screw to give additional friction.

Spring Washers Split spring washers, with or without serrations on the outer faces, may be employed to prevent a nut from loosening.

Simmond's Nut This type of nut has a fibre or nylon ring set in an annulus cut inside the nut. As the nut is tightened the ring is forced into the thread of the stud or bolt and the compression and friction give the required locking action. Frequently used on automobile construction, but should be replaced on dismantling.

Taper Pin After tightening the nut, a taper pin may be driven through a hole drilled and taper reamed to fit the pin, the ends of which are opened after insertion to prevent accidental withdrawal.

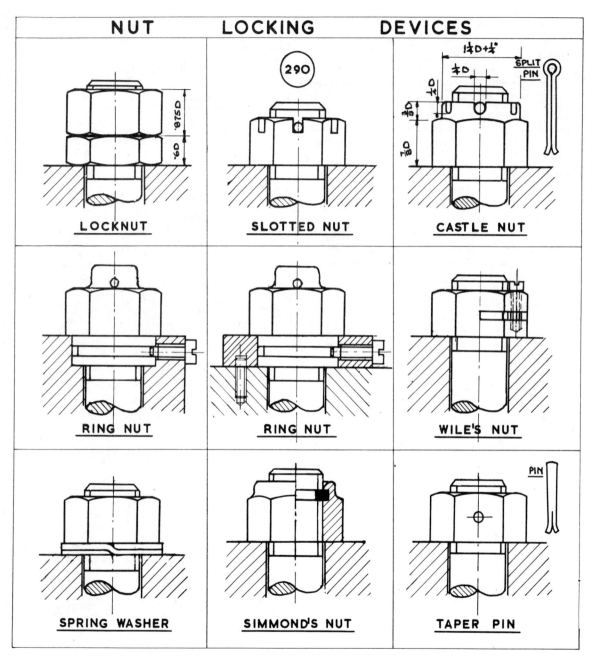

NUT LOCKING DEVICES

LOCKNUT

SLOTTED NUT

CASTLE NUT

RING NUT

RING NUT

WILE'S NUT

SPRING WASHER

SIMMOND'S NUT

TAPER PIN

291. Locking Devices Nuts and bolts may be prevented from loosening by adopting various anchoring devices.

Locking Plate This plate has a double hexagonal aperture cut in one end which allows it to be dropped over the tightened nut or bolt head. The plate is secured to the component by a set screw.

Twin Tab Washers The stamped sheet metal strip is tightened under the bolt heads and the tabs of the washer are then bent up closely to the side of the hexagonal face.

Single tab washers may be used at corners and edge locations by turning the tab over the corner of the component. When the tab washer is used in a central position, a hole is drilled in the component to receive the tab.

Wired Bolts In this method, the bolt heads are drilled to receive a wire which is passed through the holes after the bolts are tightened, the ends of the wire are twisted together to fasten off. It is essential that the direction of the wiring prevents loosening of the bolts.

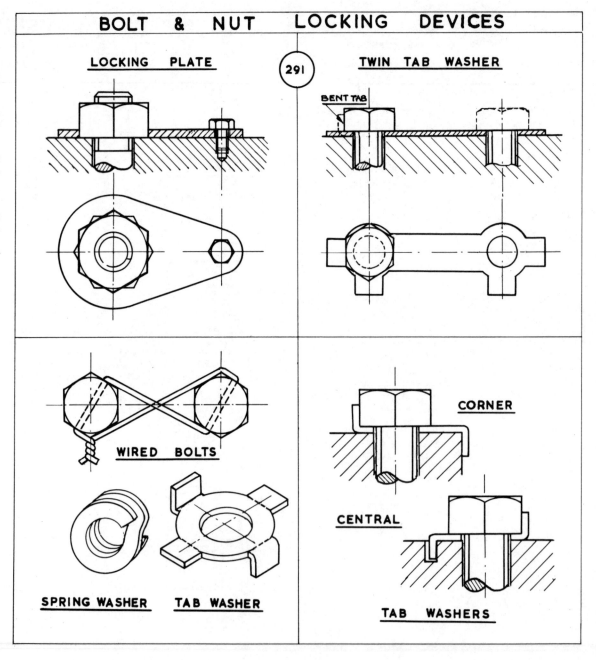

BOLT & NUT LOCKING DEVICES

LOCKING PLATE

291

TWIN TAB WASHER

BENT TAB

WIRED BOLTS

SPRING WASHER   TAB WASHER

CORNER

CENTRAL

TAB WASHERS

292. Riveted Joints

d = rivet diameter.
p = rivet pitch.
t = plate thickness.
f_t = safe tensile stress for plates.
f_s = safe tensile stress for rivets.
f_c = safe crushing stress for rivets and plates.

$\dfrac{\textbf{Ultimate strength}}{\text{Working stress}}$ = safety factor.

$$\text{Single shear} = (p-d) \times t \times f_t$$
$$= \tfrac{1}{4}\pi \times d^2 \times f_s$$
$$= d \times t \times f_c.$$
$$\text{Double shear} = (p-d)t \times t \times f_t$$
$$= \tfrac{1}{2}\pi \times d^2 \times f_s$$
$$= d \times t \times f_c.$$

Unwin's Formulae, $1\cdot2\sqrt{t}$, if t is greater than $\tfrac{5}{16}''$; if less, make shear resistance = crushing resistance.

$$t = \frac{d^2}{1\cdot44}$$

$$p = \frac{\pi d^2 f_s}{4 t f_t} + d. \qquad \begin{array}{l} d = 2\cdot55t, \text{ single shear.} \\ d = 1\cdot275t, \text{ double shear.} \end{array}$$

Strap thickness,
  Single cover, 1·25 thickness of plate.
  Double cover, 0·75 thickness of plate.

Steel Safety Stress Values,

Plates, f_t	= 12,000 lb/in^2
Single rivets, f_s	= 9,500 lb/in^2
Double rivets, f_s	= 8,750 lb/in^2
Single, f_c	= 16,000 lb/in^2
Double, f_c	= 21,000 lb/in^2

293. Exercises

1. From the diagram 281 and relevant tables in the appendix, give standard tolerances for running fit, location fit, key and press fit for a steel shaft $1\tfrac{1}{4}''$ diam. Tabulate the results.

2. From the diagram 282 give the Newall tolerances for Y running fit; push fit; driving fit; force fit for a $\tfrac{3}{4}''$ diam. steel shaft. Tables in the appendix.

3. Show the disposition and size of rivets in a double cover riveted joint fastening two steel plates 6″ wide and $\tfrac{1}{2}''$ thick.

4. Show details of a riveted joint zig-zag fastening two plates 8″ wide and $\tfrac{5}{8}''$ thick, flush finish.

5. Sketch three methods in which a riveted joint may fail. Give reasons for the failure.

6. Draw diagrams showing the riveted joint between a stanchion I section, 8″ × 6″, and a beam of similar dimensions.

7. Give a sketch of a roof truss No. 137 made from 3″ × $\tfrac{3}{8}''$ steel angle, span 30 ft, joints having gussets made from $\tfrac{3}{8}''$ plate. Show the centres only of the rivets.

8. Sketch four methods of riveting thin sheet, two giving a flush surface, one for inaccessible positions.

9. Draw a sectional view of two $\tfrac{1}{2}''$ thick steel plates held together by a suitable Whitworth nut and bolt.

10. Draw a sectional view showing two $\tfrac{1}{2}''$ thick plates held by a nut and bolt giving a flush finish to one side.

11. Draw the sectional view showing a cylinder head 2″ thick held down by a $\tfrac{1}{2}''$ diam. stud.

12. Draw a sectional view showing a 1″ thick plate fastened to a casting by suitable Whitworth set bolt.

13. A brass plate $\tfrac{1}{4}''$ thick is set screwed to a casting to give a flush finish. The plate is 4″ × $2\tfrac{1}{2}''$; show spacing and one setscrew 3 × f.s.

14. Draw a Unified screwthread section pitch 2″.

15. Draw the Standard Whitworth thread section pitch $2\tfrac{1}{2}''$.

16. Draw the Acme thread and the Buttress thread sections pitch 2″; draw three crests in the above diagrams.

17. Make a conventional drawing of a nut and bolt as in No. 289. Take D = 2″.

18. Draw any five of the various bolts or screws shown in No. 291. Take D = 1″.

19. Draw any three methods of nut locking devices shown in No. 292. Take D = $1\tfrac{1}{4}''$.

20. Draw sectional views of three of the locking devices shown in No. 293. Take D = 1″.

[handwritten note: Draw sectional views of two of a locking surfaces]

SHAFTS, BEARINGS AND SEALS

6

Keys and Keyways

Cottered Joints

Shaft Couplings
Rigid and Flexible

Universal Couplings

Clutches

Bearings

Seals

294. Keys and Keyways Keys are used to fix shafts and wheels and collars together, either rigidly or with limited axial movement.

Rectangular Key A rectangular recess is cut in both wheel and shaft to the proportions shown, and the key fitted.

Gib Head Key This tapered key provides a rigid fixing when driven into position. The gib head enables the key to be withdrawn for dismantling.

Feather Key The parallel-sided key is setscrewed into an end milled slot in the shaft. This key allows axial movement of the wheel on the shaft while still transmitting rotary movement. It also facilitates assembly of parts under difficult conditions.

Woodruff Key A segmental shaped key which fits into a corresponding recess cut by a standard size cutter. This key is used on tapers and other positions where self aligning is required.

Saddle Keys Used for temporary fixing or light loads.

Round Keys These may either be screwed or tapered. The diagram shows a screwed pin, the head of which is sawn off after insertion to give a flush finish.

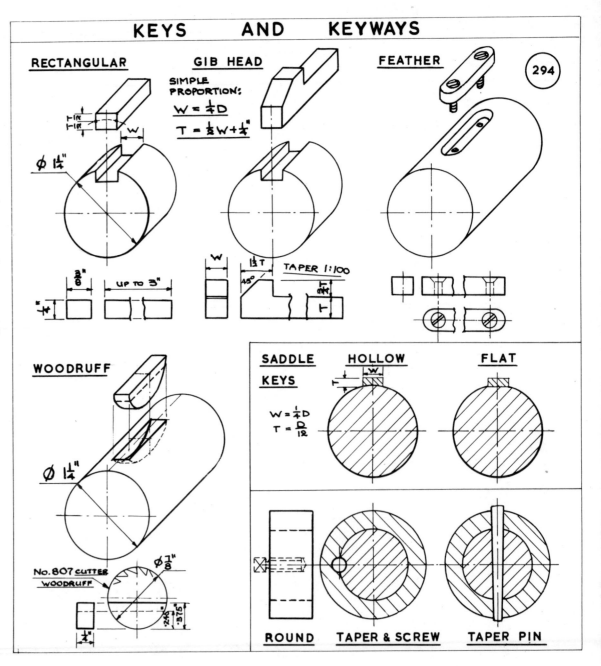

295. Splined Shafts A shaft may be grooved to form splines or keys enabling gears or similar components to slide axially whilst still transmitting rotary motion. This method is adopted in gearboxes when the gear shift lever engages or disengages a gear by means of the shift claw working in an annulus as shown in the diagram.

296. Split Tapered Cotters are used to retain the collar for the spring on poppet valves in the internal combustion engine.

297. A Tapered Cotter tightened by nut and washer can be used for locking a shaft and lever or crank together. The cotter requires a flat machined on the shaft.

298. A Flat Tapered Cotter is used for locking a shaft and base or wheel together. Simple proportions are given for general use, a taper of 1 : 30 is usually adopted.

299. Cottered Joint for Rods Two rods may be joined by a tapered spigot joint locked by a flat tapered cotter. The proportions shown give a joint of uniform strength in terms of D = diameter of rod.

300. Cotter and Gib for strap end of engine connecting rod. The bearing brasses of the big end of steam engines are connected to the connecting rod by means of a strap which in turn is locked to the rod with a gib and cotter. The jaws of the gib prevent the strap from opening as the cotter is driven home, and since its outer edge is tapered as the cotter, a parallel wedging action takes place.

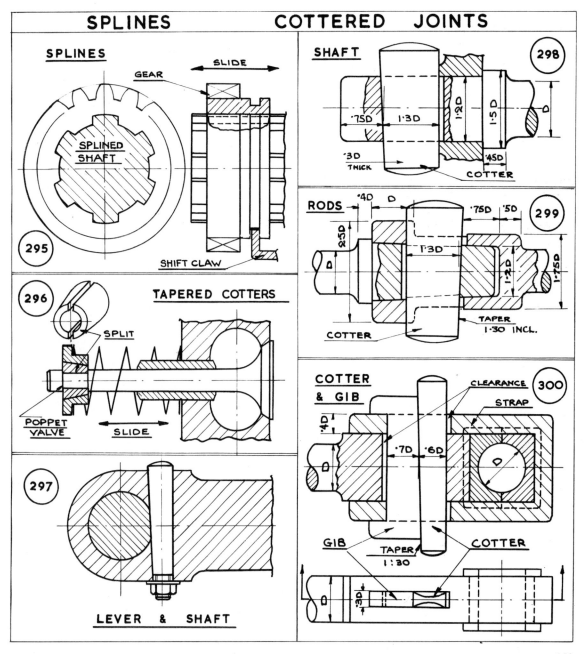

SPLINES   COTTERED JOINTS

159

301. Muff Coupling for Shafts Two shafts of equal diameter may be rigidly connected by means of a muff coupling. This consists of a cylinder machined in two halves and secured on the shafts by nuts and bolts. The shafts and muff are slotted for a long rectangular key. The general proportions of the muff are shown in terms of the diameter of the shaft.

302. Flanged Coupling Two shafts may be rigidly coupled by using flanged coupling plates. The couplings are machined with matching turned projection and annulus for location and are press fits on the shafts. The shafts and couplings are slotted for taper keys driven in from the meeting faces. The flanges and shafts are finally bolted together the bolts being tight fits in the flange holes.

   The general proportions are given in terms of the shaft diameter. See tables.

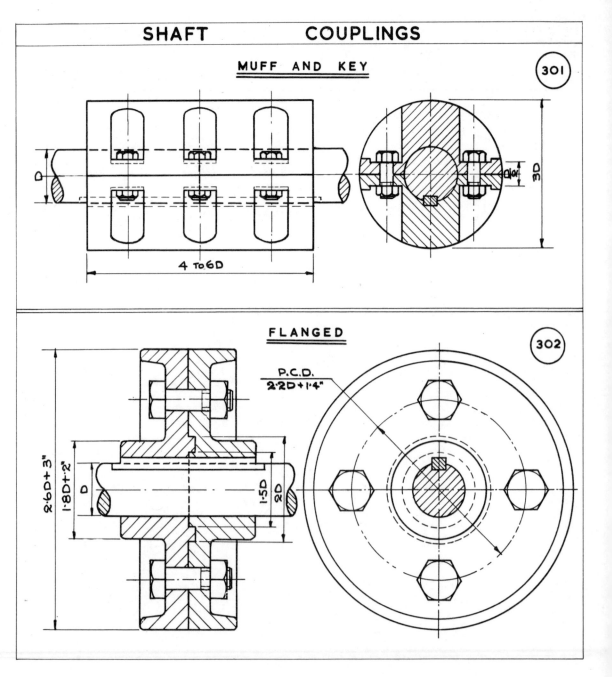

SHAFT   COUPLINGS

MUFF AND KEY   301

4 TO 6D

3D

FLANGED   302

P.C.D.
2·2D+1·4"

2·6D+3"   1·8D+2"   D   1·5D   2D

VARIOUS COUPLINGS

303. Flexible Coupling Moulded Rubber Insert. Where two shafts are to be connected and some slight mal-alignment and torsional strain is anticipated, a suitable flexible coupling is obtained by the use of a moulded rubber insert between the coupling dogs. The metal flanges are keyed to the shafts and have projections which fit into the moulded recesses in the rubber insert, through which the drive is transmitted and shocks absorbed.

304. Bonded Steel and Rubber Flexible Coupling. This coupling employs a flexible rubber disc bonded to two steel plates which are bolted to flanges keyed to the shafts. The bolt holes are arranged alternately and irregularities of transmission are absorbed by the disc. An Isometric View is given later.

305. Rigid Compression Coupling The diagram shows a compression coupling in which the coned flanges compress a mating double conical sleeve which is slit to allow it to be reduced radially. This type of coupling grips by friction and uses no keys; it is capable of being dismantled without disturbing the shafts.

306. Chain Coupling Flexible. The flanges have chain sprocket wheels cut at the meeting faces, and are coupled by a length of duplex chain by which the drive is transmitted. The flanges are keyed to the shafts, and the coupling is shrouded by a cover. The chain allows for some flexibility in mal-alignment of the shafts, and the drive can be dismantled by the removal of the cover and chain.

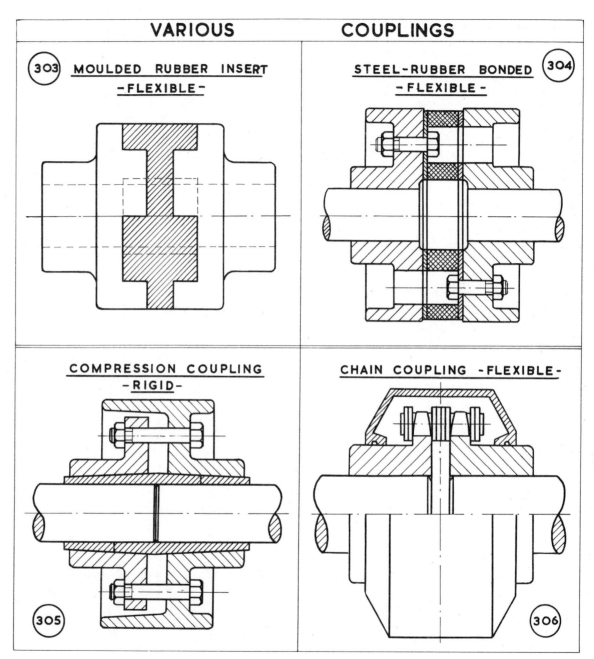

VARIOUS COUPLINGS

303 MOULDED RUBBER INSERT -FLEXIBLE-

STEEL-RUBBER BONDED 304 -FLEXIBLE-

COMPRESSION COUPLING -RIGID- 305

CHAIN COUPLING -FLEXIBLE- 306

UNIVERSAL JOINT COUPLINGS

Where drive has to be transmitted through shafts having an angularity of more than 5°, a universal joint coupling is employed. This usually consists of two yokes and a cross trunnion piece.

307. Hooke's Type Universal Joint The diagrams show a typical Hooke's joint consisting of two yokes keyed to the ends of the shafts, and connected to a spider or cross in which screwed pins form bearings for the yokes. Proportions are given in terms of the diameter of the shafts. This joint will transmit drive through 20° of angularity.

308. Automobile Type Universal Joint A universal coupling used in transmitting drive from a flexibly mounted engine to the sprung back axle unit of an automobile. The coupling has to work under varying loads often at high speeds and through rapidly changing angularity. Needle roller bearings are used on the trunnions, held in position by spring circlips. Splined shafts are employed to allow extension of the shaft since a universal joint is used at both ends of the transmission shaft and a double angularity is often experienced.

162

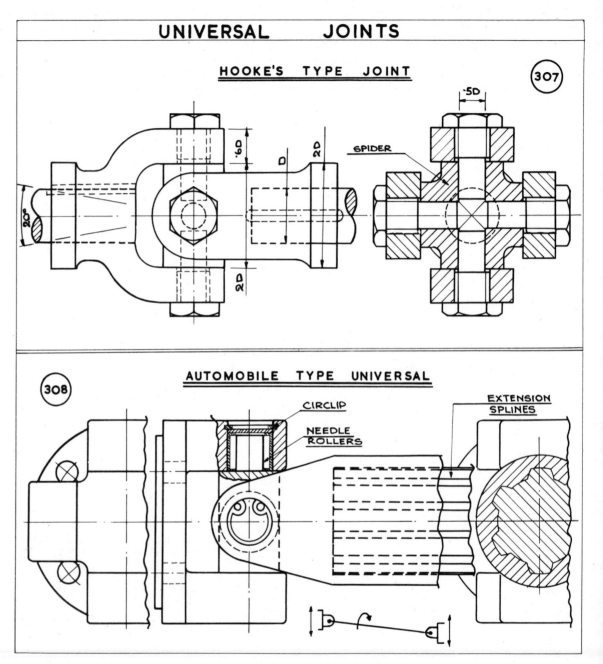

CLUTCHES

Where a drive needs to be completely disengaged and re-engaged frequently, some form of clutch must be introduced. One side of the coupling is moved on a splined shaft by means of a foot or hand operated lever often returned by a spring.

309. Claw or Dog Clutch One side of the coupling is keyed permanently to one shaft. The other coupling slides on a splined shaft or on a feather key and is operated by a lever. Mating projections on the flanges transmit the drive, and this simple type of clutch can only be re-engaged at low speeds.

310. Cone Clutch The clutch has two mating conical surfaces, one surface may be coated with leather or bonded lining. The cones allow easy disengagement and re-engagement of the drive from rest position, the drive is by friction between the two engaging surfaces. An Isometric View is shown later.

311. Dry Plate Clutch This type of clutch is used in many automobiles, and consists of a light steel plate with asbestos friction lining rings compressed between the flywheel surface and the driven presser plate connected to the gearbox. On depressing the clutch foot-pedal, the plate is freed and the drive lost, being automatically re-engaged as the foot pedal is released. A carbon pressure ring or ball bearing takes the thrust of the foot-pedal, the surface pressure on the plate linings is usually between 15 and 30 lb/in².

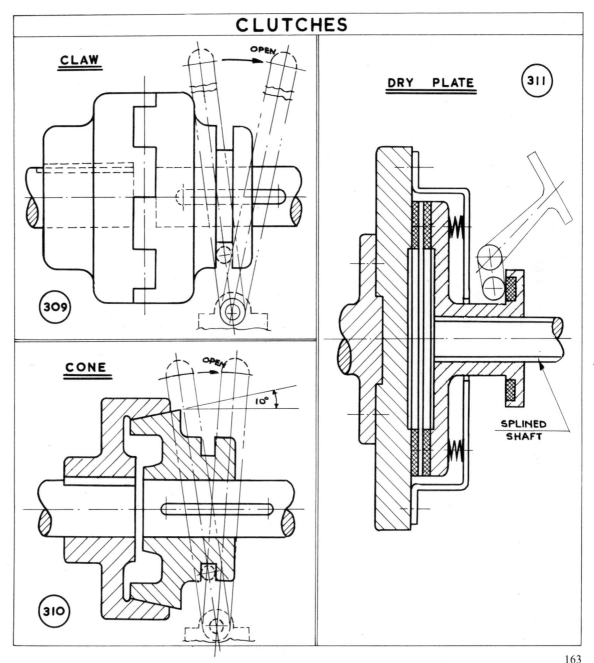

163

SHAFT BEARINGS

Shafts and axles need supports lined with anti-friction surfaces and provision for lubrication.

The support is usually an iron casting or forging either in one piece or split with inset bearing surfaces of bronze, gunmetal, white metal, lead bronze, sintered metals, and plastics, depending upon the use and conditions.

312. Simple Bearing   A simple bearing with a pressed bronze bush, oil hole drilled in top surface.

313. Halved Bearing   The support is made in two parts bolted together, the 'brasses' are also split and located by flanges.

314. Footstep Bearing   Used for the lower end of a vertical shaft. A hardened steel plate takes the weight of the shaft. A ball thrust race may alternatively be used, and is shown in a later diagram.

315. Ring Oiler Bearing   A loose ring rolled by friction from the shaft dips into an oil well and continually lubricates the bearing.

Other methods of simple lubrication are by drip feed, wick feed and felt oilsoaked pads.

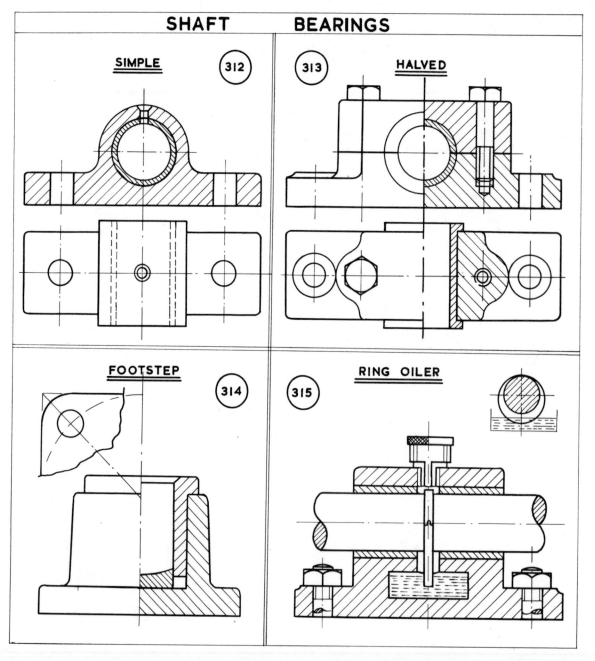

BALL BEARINGS

316. Single Row Ball Bearing The race is composed of an inner and outer hardened grooved ring between which run hardened steel balls carried and spaced in a cage. Ball bearings are made in a wide range of sizes, the smallest are found in watches whilst the largest are used in turrets of heavy naval guns.

The simple type shown usually fits into a recess turned in the housing and is retained by a collar or by shoulders on the shaft.

317. Self-aligning Ball Bearing The outer shell is ground to the same spherical line as the housing enabling the bearing to align itself to the shaft line which may vary as the load is applied. The bearing also adjusts itself to small inexactitudes arising if the housing is incorrectly laid.

318. Thrust Race Ball bearings may be used to take end thrust as shown in the footstep bearing for a vertical shaft. The axial thrust is taken by the balls carried in a cage and moving in circular grooves ground in hardened steel plates.

319. Spring-loaded Ball Bearing Where some axial movement of the shaft may take place due to expansion or load, the race may be held to the shaft by a screwed collar and the bearing moves against a spring held in the housing.

320. Ball Races require lubrication to avoid becoming seized under load. The housing may be arranged to act as an oil bath requiring sealing—shown later—or the race may be packed with grease sealed in with spring fit plates as shown, the lubricant lasting the life of the bearing.

Ball races may be single or double row, designed to take radial and/or axial thrust, and since the balls are in point contact with the shells much less friction is generated than with the simple journal bearing.

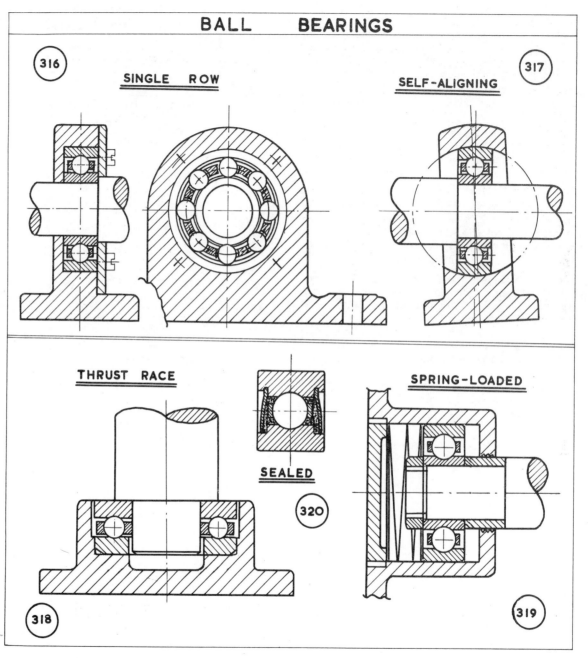

BALL BEARINGS

316 — SINGLE ROW
317 — SELF-ALIGNING
318 — THRUST RACE
320 — SEALED
319 — SPRING-LOADED

321. Roller Bearings In bearings which carry heavier loads, roller bearings are used, the line contact of the roller and shell giving better radial capacity. Rollers are spaced in cages and may be arranged to run in a groove cut in the inner shell whilst the outer shell being plain, allows of some axial movement. The rollers may be arranged in single, double or triple rows to carry greater loads.

322. Needle Roller Bearings Caged needle roller bearings can be used where space is restricted. The end of the roller is reduced and rounded to prevent it acting as an end mill and cutting the cage. The rollers run on the shaft which requires to be hardened.

323. Sealed Roller Bearing The diagram shows a heavy duty roller bearing. The inner shell is secured to the shaft by a screwed collar and tab washer. The housing is sealed by synthetic rubber or plastic rings set in grooves. The housing acts as an oil reservoir.

324. Tapered Roller Bearing Where radial and axial thrust is present, tapered rollers can be used. The diagram shows a bearing with double tapered rollers used in automobile front wheels capable of absorbing radial and double axial thrust at high speeds. The bearing is lithium grease packed and sealed with a garter spring seal, the bearings are secured and adjusted by a slotted nut and split pin.

Shafts may employ both ball bearings and roller bearings. The ball bearing at one end may be used to absorb axial movement, and the roller bearing at the second end to bear the heavier load.

Other types of seals are shown later.
Refer to B.S. 292.

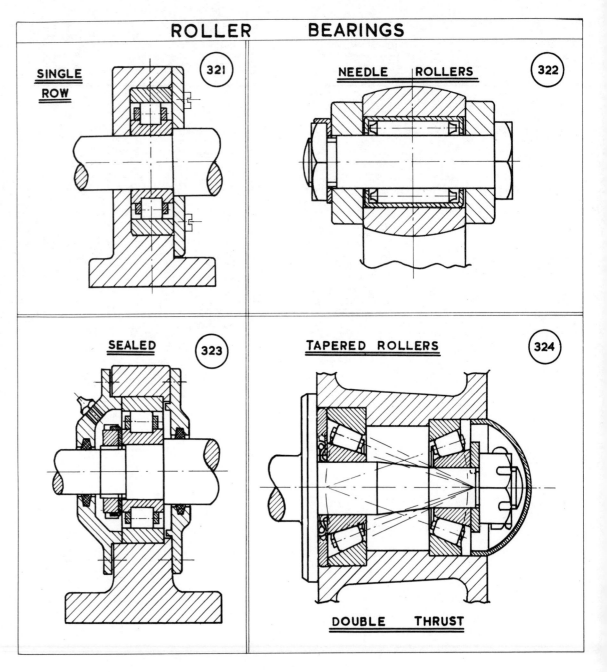

ROLLER   BEARINGS

SINGLE ROW — 321

NEEDLE   ROLLERS — 322

SEALED — 323

TAPERED ROLLERS — 324

DOUBLE   THRUST

SEALS

Shafts and rams in hydraulic systems often require high pressure sealing against leakage. This is accomplished by a sealing ring forced against both shaft and housing. Early types of seals, leather, braided hemp, cotton, flax, have largely been superseded by synthetic rubber, butyl, silicone, nylon and other plastics. These materials stand up to modern solvents.

325. Labyrinth Seals   The seal and gland shown, consists of a number of rings of vee section, spring loaded. The rings may be of metal and plastic, or metal and metal. White metal and a bronze are one combination.

326. Split Ring Seal   The seal rings are packed in sections inside cups, and a garter spring applies concentric pressure closing the seals on the shaft and to the housing.

A number of units may be used in line, held in place by a keeper plate and bolts. The units may be replaced without disturbing the shaft.

327. 'U' Seal   A moulded U section seal of leather or synthetic material is held in position by the gland. The pressure of the pumped liquid tends to force open the U seal and improves the sealing.

328. Garter Spring   A moulded leather or synthetic material seal held in place by a garter spring. Used for light pressures, automobile shafts and gearboxes.

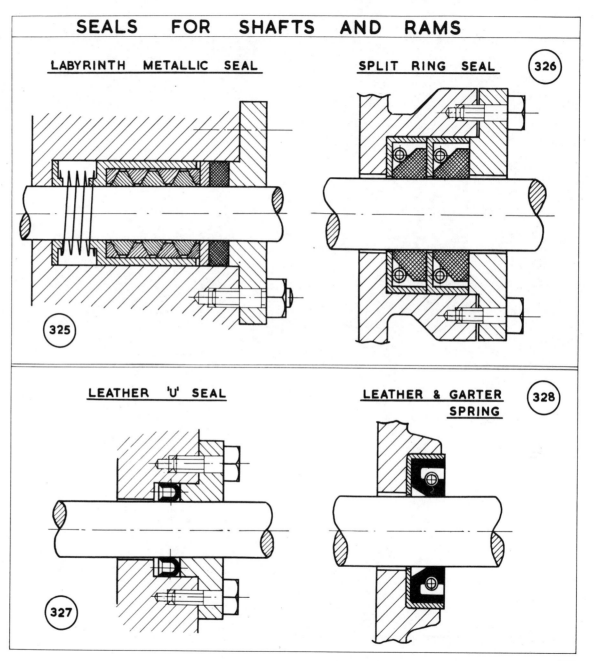

SEALS FOR SHAFTS AND RAMS

LABYRINTH METALLIC SEAL

SPLIT RING SEAL   326

325

LEATHER 'U' SEAL

LEATHER & GARTER SPRING   328

327

PIPES, JOINTS AND VALVES

Pipes and Couplings

Expansion Joints

Non-return Valves

Safety Valves

Stop Valves

329. Pipe fittings A selection of standard pipe fittings are shown, sizes are given in the example for pipes of a nominal bore of 1″. The ends of the pipes are threaded (1″ bore to 6″ bore all use 11 threads per inch), and a socket piece is used to join straight lengths. An end may be sealed by a cap fitting. Tee junctions enable pipes to be joined at right angles. Elbow pieces may be right angles or 45°. Pipes may also be joined by flanged bends which are bolted together. Standard sizes and proportions are listed in detail in the B.S. Handbooks.

B.S. 10:1962. Pipe flanges.
B.S. 27. Pipe threads.
B.S. 1256. Copper pipe fittings.
B.S. 143. C.I. and copper pipe fittings.
B.S. 1740. Wrot pipe fittings B.S.P. thread.

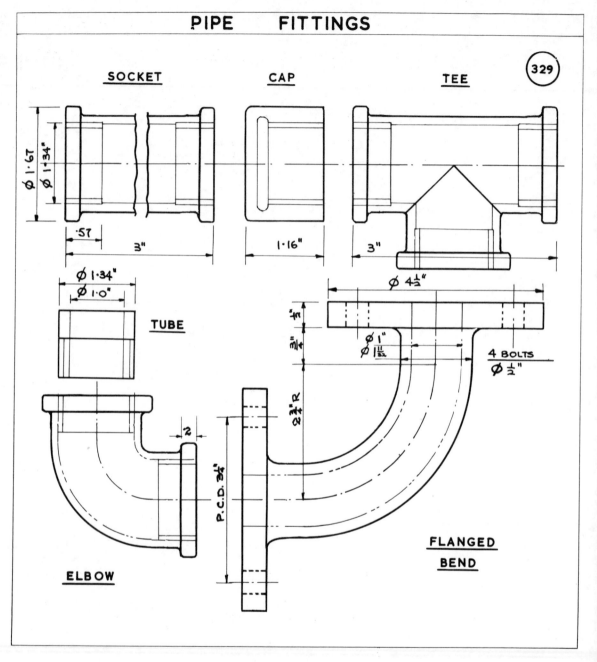

170

330. Flanged Pipe Couplings Pipes may be joined by screwing flanges to the pipe ends and bolting the sections together. A taper thread is cut on the pipe and flange. The diagram shows details of flanges suitable for liquid and gas pressures up to 50 lb/in², and on pipe of nominal 1″ bore.

Heavy Duty Flange The diagram shows details of a coupling capable of withstanding higher pressures up to 450 lb/in². The pipe and flange are taper screwed together, the flange is heavier, and the bolts greater in diameter.

Screwed and Welded Flanges In this type, the heavy flange is screwed on a taper thread and a vee-cut is welded up. The joint is machined to fit, and the sections bolted together.

331. Hydraulic Couplings Extra high liquid pressure systems using pressures of 1,500 lb/in² require special couplings. The heavy flanges are screwed on the pipes, one pipe penetrating into the opposite flange, the seal being made by compression on a soft copper ring. The flange shape is shown, and has two bolts.

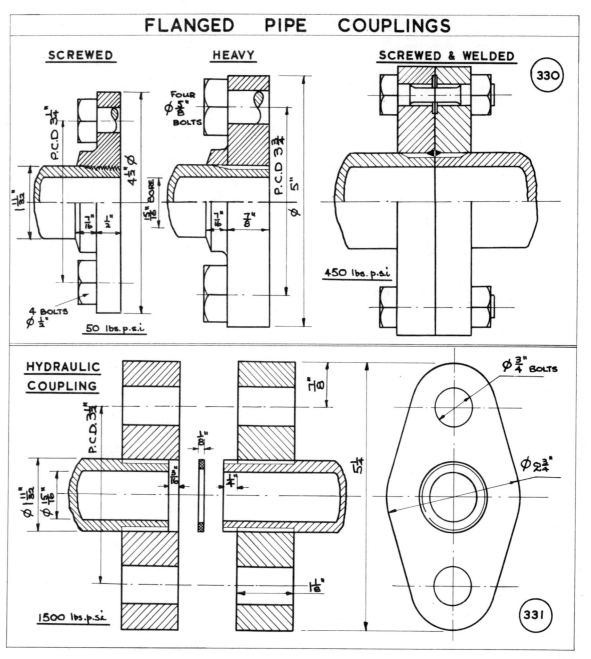

171

332. Hydraulic Pipe Couplings A further coupling is shown having square flanges. The sealing is effected as before by the compression of a soft copper ring between the ends of the meeting pipes. Sizes are given for a pipe of nominal 1″ bore, pressures up to 2,250 lb/in².

SMALL BORE HYDRAULIC JOINTS

333. Coned Union A separate double coned sleeve is placed on the tube and is compressed when the union nut is tightened. The coned sleeve may be soldered or brazed to the tube.

334. Nut and Sleeve An in line joint between two tubes can be effected by the use of an internal coned sleeve and outer ring which compresses the end of the tube to a seal when the union nut is tightened.

335. Brazed and Screwed Sleeves are hard soldered to the tubes, the solder having being contained in the annular groove in the sleeve. The flanges are screwed to the sleeves. The joint seal is effected by compressing a joint ring when bolting up the flanges. This type of coupling is used on thinner walled tubing.

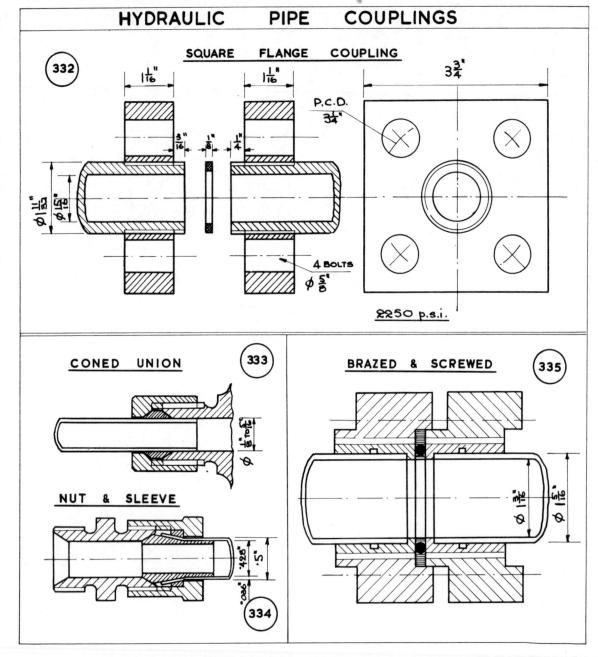

336. Expansion Joints and Bends Expansion of pipes carrying steam or hot liquids require special provision and methods adopted are lyre bends and the U bends in the pipe line. The bends may be made up of bends built to the form shown.

337. Expansion Joint The unit consists of two tubes one of which slides inside the other, the gland sealing the joint whilst allowing axial movement. The movement is limited by the length of the slide bars. The seal ring is asbestos fibre. The joint is drawn suitable for pipes of 1″ nominal bore.

EXPANSION JOINT AND BENDS

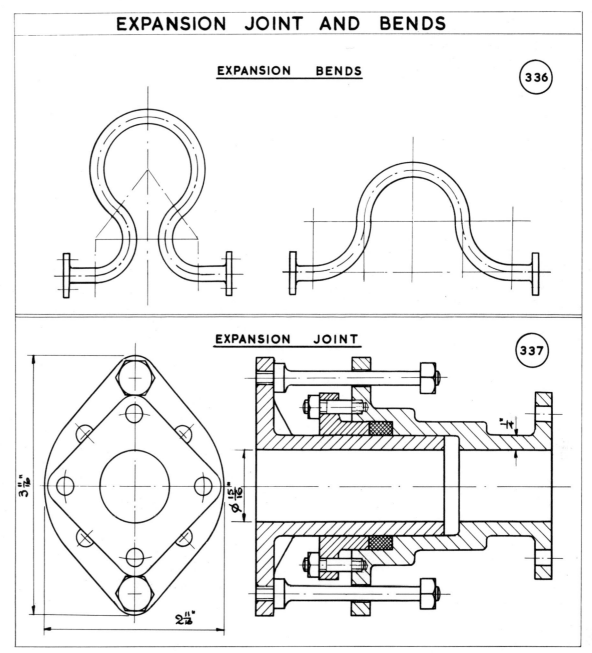

EXPANSION BENDS

336

EXPANSION JOINT

337

NON-RETURN VALVES

To prevent the back flow of gases and liquids in systems a number of types of non-return valves are employed.

338. The Poppet mushroom headed valve is adopted in most internal combustion engines for inlet and exhaust purposes. The valve is opened mechanically by a cam, and returned by the spring to its seat after a measured lift, dwell and fall, generated by the cam shape.

339. Release Valves The diagram shows a spring-loaded needle valve seating on a circular seating. The flow lifts the valve on reaching a pressure exceeding that of the pre-set loading of the valve. This type is used on hydraulic systems.

340. Feed Check Valves This type of valve is used in the flow of boiler feed water and uses a disc valve with three or four aligning vanes to restore the non-return valve to its seating, helped by a light spring.

341. Ball Check Feed Valve This type of non-return valve employs a bronze or stainless steel ball as the valve. Plastic balls may be used where the pressure is low as in cooling liquid systems.

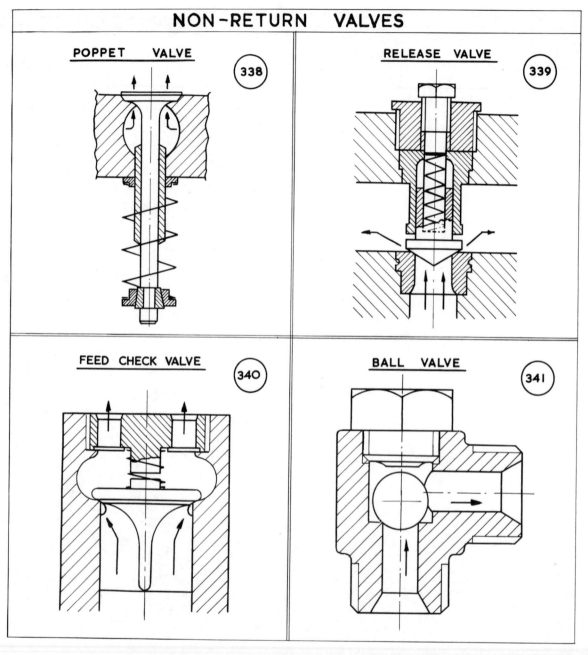

NON-RETURN VALVES

POPPET VALVE 338

RELEASE VALVE 339

FEED CHECK VALVE 340

BALL VALVE 341

342. Non-Return Box Flap Valve An in-line gravity fall non-return valve for use in gas pipes. The flap valve is pivoted at the top edge and closes the pipe if an explosive blow-back ignites the gas beyond the valve. It is used as a safety measure in gas systems. In normal use the flap valve is opened by the flow of gas. The sealing face is of asbestos lining.

343. Ball Release Valve In hydraulic and steam or gas systems a ball valve may be spring loaded to lift off and release excess pressure at a safe limit.

344. Release Valve Spherical ended plunger, spring loaded. The release valve is similar to the previous example, the valve being a plunger, hollow, containing the spring. The cap is bolted down. The spherical ended plunger requires relief holes drilled in it above the seal line to avoid pressure build-up in the spring tube, which could interfere with the normal lift-action of the valve.

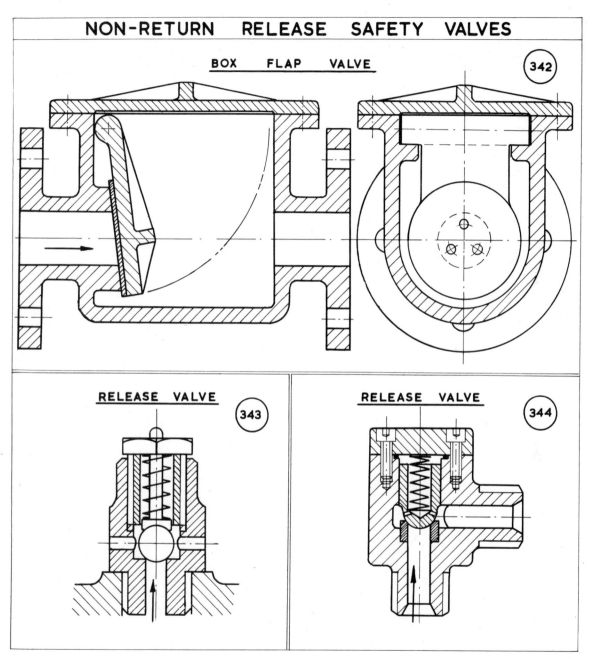

NON-RETURN RELEASE SAFETY VALVES

BOX FLAP VALVE   342

RELEASE VALVE   343

RELEASE VALVE   344

345. Disc Non-Return Valve A spring-loaded face valve, plastic face backed by sheet metal. The valve is used in lift pumps.

346. Line Pressure Valve A spring-loaded valve set to open at a fixed pressure. The union joints are made by a seal ring which enters a groove in the pipe end.

347. Steam Safety Valves A gunmetal valve is held on its seating by a lever and weight. The valve is vaned to direct the valve during re-seating.

The safety valve is bolted to the boiler top and is set to lift at a safe pressure.

348. Ramsbottom Safety Valve Two vaned valves are used, held down by a common lever which is held down by a spring and slotted link. Both valves can be manually tested at set intervals by easing the lever up and down. Used on steam locomotives.

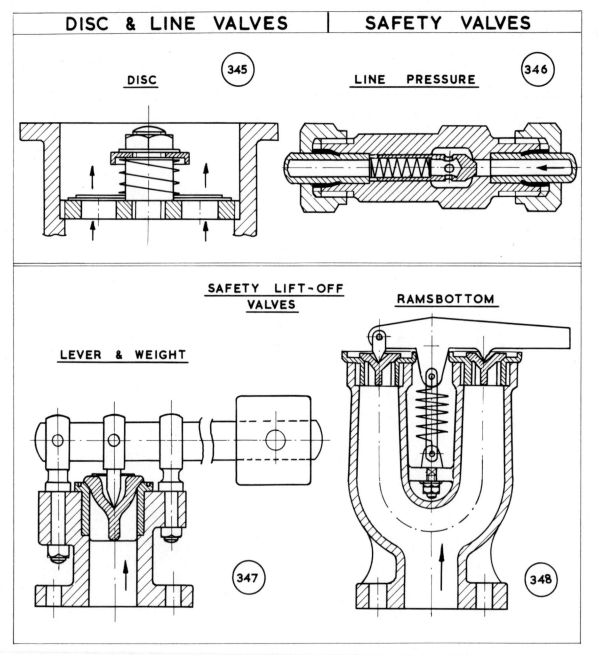

DISC & LINE VALVES | SAFETY VALVES

DISC (345)

LINE PRESSURE (346)

SAFETY LIFT-OFF VALVES

LEVER & WEIGHT (347)

RAMSBOTTOM (348)

STOP VALVES

Screw down valves are used for cutting off a supply in hydraulic steam and gas systems.

349. Needle Screw Down Valve A 45° pointed screw plunger closes on a seating to shut down a supply. A screw down gland and sealing ring make the plunger gas tight. The inlet and outlet pipes are connected by coned unions. Suitable for high pressure small bore systems up to $\frac{3}{8}''$ diam., and low pressure to $\frac{3}{4}''$ bore.

350. Screw Down Valve The diagram shows a standard type screw shut down valve flanged for bolting into pipe systems.

The body is a casting. The spindle screws down through a bridge carried by the top plate. A gland makes the spindle gas tight and the sealing ring is compressed by the two bolts of the gland. The valve seats on the centre cast web, and is located by a pin. The valve has an open-sided slot enabling it to be fitted on the end of the spindle.

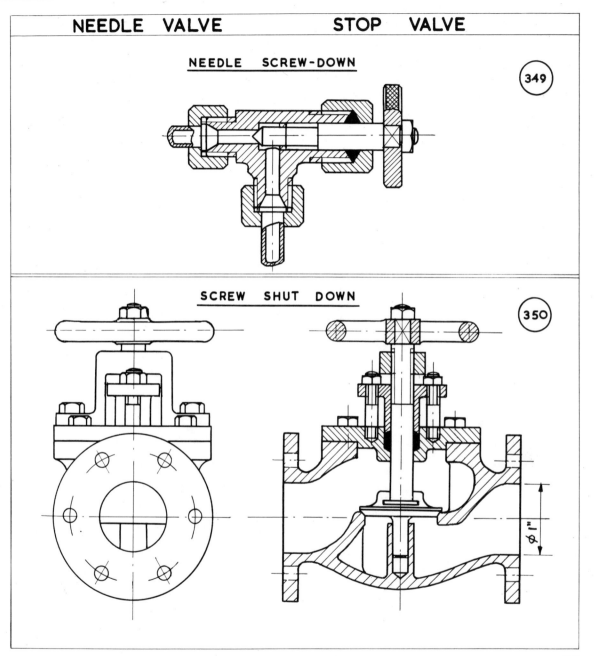

NEEDLE VALVE   STOP VALVE

NEEDLE SCREW-DOWN

349

SCREW SHUT DOWN

350

177

351. Needle Line Valve The diagram shows a needle valve for an in-line joint. The conical ended plunger screws down to seat. A gland holds a sealing ring. The pipes are attached by coned union joints.

351A. Heavy Duty Hydraulic Valve The screw down stop valve has a heavy cast body and stout spindle which works in a screwed collar. The collar and gland are secured by bolted flanges. The oval flanges carry two heavy bolts.

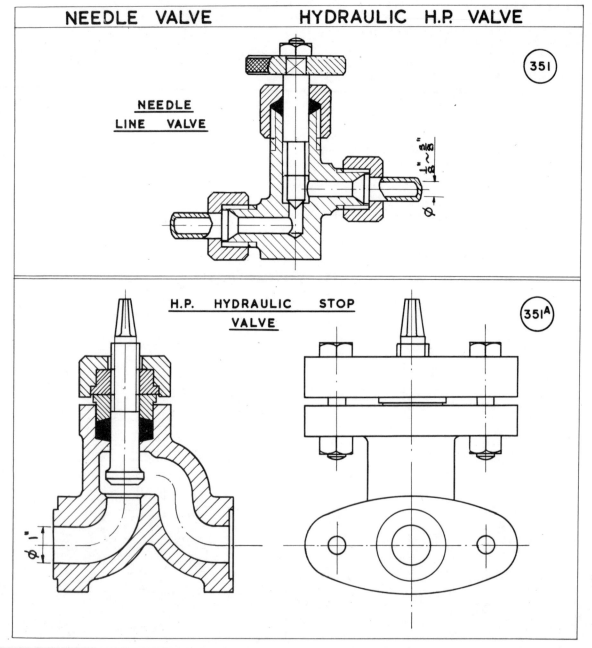

NEEDLE
LINE VALVE

351

H.P. HYDRAULIC STOP
VALVE

351A

PUMPS

Ram. Oscillating. Centrifugal

Gear. Diaphragm. Disc Valve

Screw. Cam and Plunger

ENGINE PARTS

Piston. Crosshead

Connecting Rod

Crankshaft

Eccentrics

I.C. ENGINES

Two Stroke

Four Stroke

PARTS

Carburettors

Ignition

PUMPS

352. Ram and Eccentric A cylindrical ram worked by an eccentric crank draws in a charge lifting the lower valve and closing the upper ball valve. The delivery stroke closes the lower valve and the upper valve opens to pass the charge. The ram is sealed by a screwed gland and seal ring.

353. Oscillating Cylinder The oscillating cylinder is carried on a trunnion or shaft. As the cylinder oscillates, the single hole port in the cylinder coincides with the port in the standard. The diagram shows the plunger halfway down on the delivery stroke.

354. Centrifugal Vane Pump The offset rotor carries twelve sliding vanes which keep contact with the outer casing by centrifugal action. This type of pump can be used as a circulatory pump or a pressure pump.

355. Gear Pump Two equal spur gears working in a housing can be used as a pump. Used in automobile engines as oil circulatory pumps for lubricating bearings and pistons, pressure 20 to 40 lb/in^2.

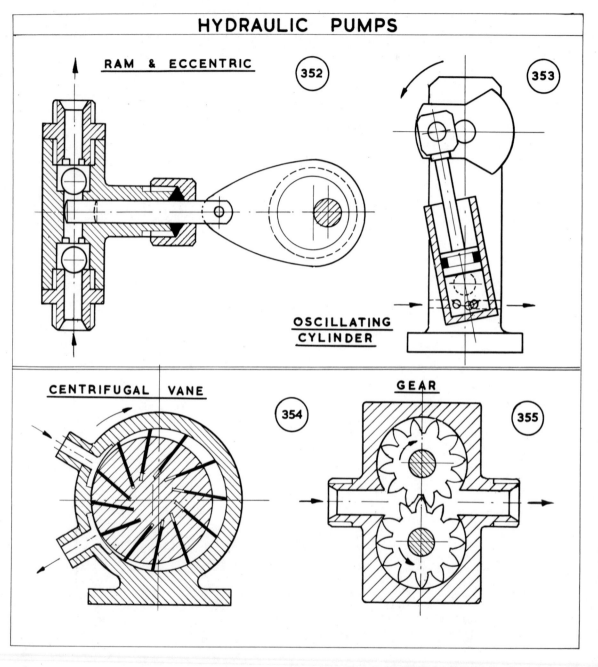

HYDRAULIC PUMPS

RAM & ECCENTRIC — 352

353

OSCILLATING CYLINDER

CENTRIFUGAL VANE — 354

GEAR — 355

356. Automobile Petrol Type Pump The pump shown employs a plastic diaphragm worked by a cam and lever from the engine camshaft. Metal or plastic disc valves, kept against their seats by light springs, control the flow of fuel. A wire mesh filter is included in the top casing. The pump has to work under light pressures to keep the carburetter level constant against the float and needle valve (shown later).

357. Disc Valve Pump A vertical spindle lifts a plunger which has a face disc valve on its upper face. A lower face valve located by a centre pin allows liquid to be drawn in as the plunger is lifted, and on the downward delivery stroke, the lower non-return valve closes and the upper valve opens to allow delivery. This is the same principle used in the century-old well pumps.

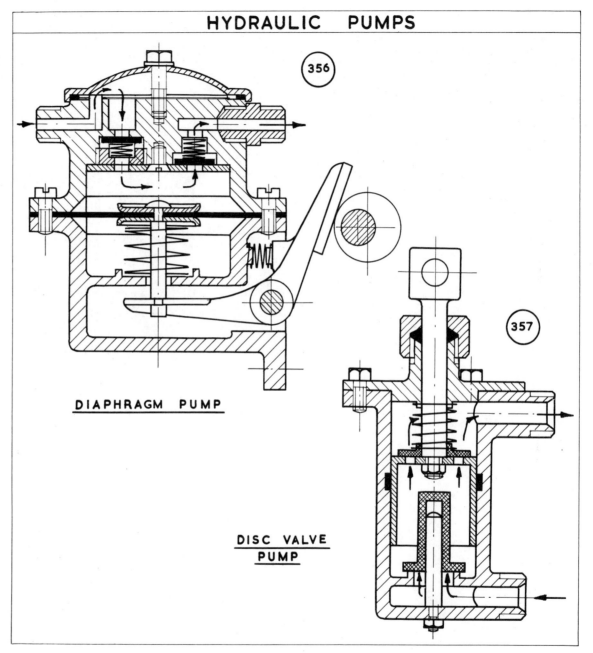

HYDRAULIC PUMPS

356

DIAPHRAGM PUMP

357

DISC VALVE
PUMP

HYDRAULIC PUMPS

358. Screw Pumps Two spindles, geared together, have left- and right-hand square threads intermeshing. Liquid is drawn in in the lower orifices, is passed along the upper and lower helical grooves of the square thread to the upper delivery opening. A seal must be included in the drive shaft and this is shown in a later diagram.

359. Plunger and Eccentric Roller bearing eccentrics drive a plunger (spring returned) and spring loaded poppet valves, non-return, control the flow of the liquid. Since the stroke is limited and short, a number of plungers are worked from the same camshaft, two, four or six units arranged to give continuous flow at high pressures.

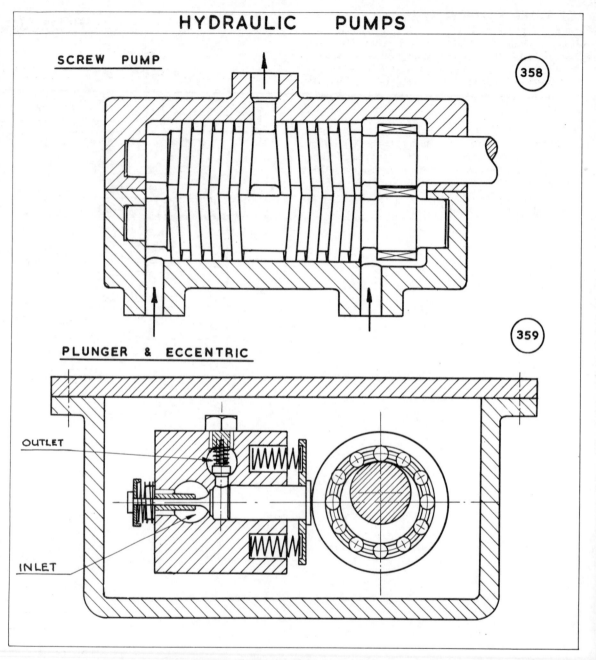

STEAM ENGINE PARTS

In steam reciprocating engines, the steam entry to the cylinder is controlled by either a slide valve or a piston valve. The steam pressure on the piston slides the piston in the cylinder and the piston rod actuates the crosshead. The connecting rod then turns the crankshaft on which are shrunk the wheels, the reciprocating action of the piston thus being turned into rotary movement.

360. Piston and Rod The cast-iron or steel piston has cast-iron piston rings set in it, and is a coned fit on the piston rod secured by a cottered nut. Cone taper 1:6.

Crosshead and Shoes The crosshead is usually a steel casting, with shoes on pins which allow for alignment. The piston rod is coned and cottered to the crosshead by a flat tapered cotter. The little end of the connecting rod works on the crosshead pin which is taper socketed into the cheeks of the crosshead, secured by a cottered nut.

361. Connecting Rod The connecting rod is a forging, the little end bronze bushed. The big end has a strap holding flanged bronze split bushes, bolted and cottered with a compensating wedge taking up the slope of the cotter. Two bolts fix the cotter. In some cases the little end is strapped, bolted and cottered to hold the bearing brasses.

The H.P. of a steam engine is given by $P.L.A.N/33,000$ where $P = lb/in^2$; $L =$ length of stroke in feet; $A =$ area of piston in in^2; $N =$ number of strokes per min.

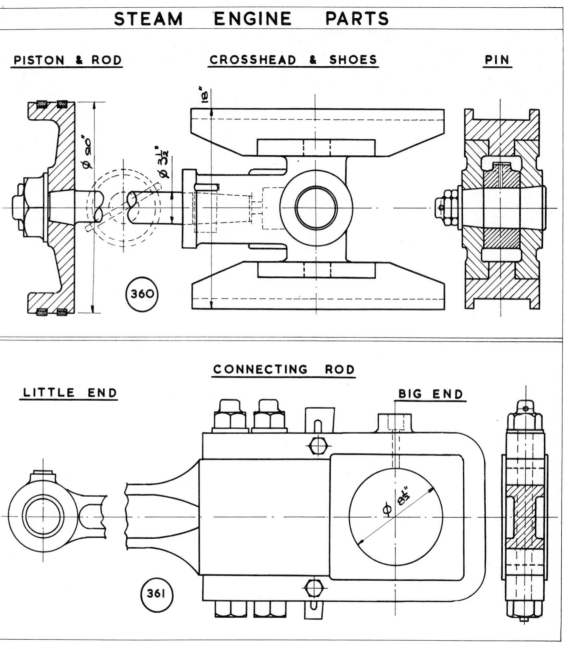

STEAM ENGINE PARTS

PISTON & ROD   CROSSHEAD & SHOES   PIN

360

CONNECTING ROD

LITTLE END   BIG END

361

183

CRANKSHAFTS

Crankshafts consist of crankpin, web(s) and mainshaft.

362. An Overhung Crankshaft has one web, forged or cast, with crankpin and mainshaft shrunk and pinned in position. It can also be forged from solid steel and machined to finished form.

The web may be extended and shaped to balance the revolving weights of the crankpin and big-end of the connecting rod and the reciprocating weights of the little-end and the piston. A simple rule is 'balance all the revolving weights and half the reciprocating weights'.

363. Locomotive Type Crankshaft A built-up, closed crankshaft is shown, with integral balance weighting formed in the webs. The crankshaft could also be forged in steel, and machined to final form.

Crankshafts may have several 'throws' equal in number to the connecting rods and cylinders.

Some internal combustion engines use a hollow cast steel crankshaft, giving strength and rigidity with lightness.

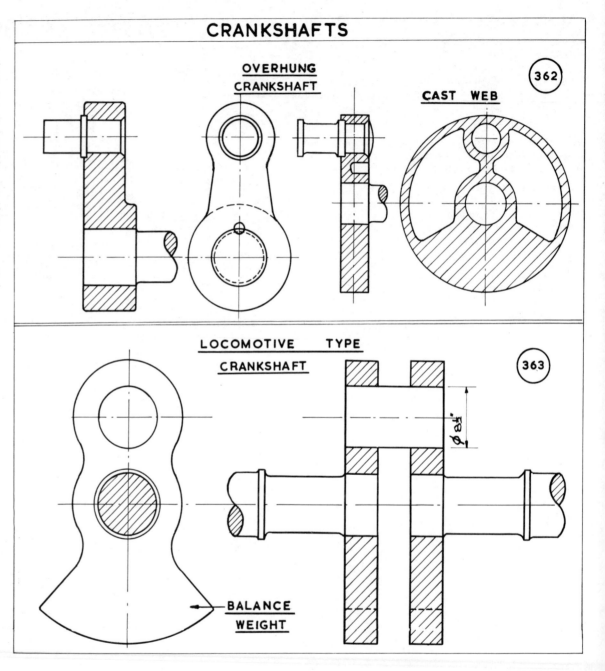

CRANKSHAFTS

OVERHUNG CRANKSHAFT

CAST WEB

362

LOCOMOTIVE TYPE CRANKSHAFT

363

BALANCE WEIGHT

364. Eccentrics for Steam Engines The slide or piston valve of the steam engine is actuated either by an eccentric crank on the mainshaft or by an overhung crank from an outer crankpin.

Eccentric and Strap The eccentric may be a casting or solid, and may be split to allow it to be erected on the shaft. It may be keyed and cottered to the shaft. The strap is made in two halves and is grooved for retention, and bolted together.

Eccentric Rod The eccentric rod is a forging and a palmate end is bolted to the strap. The little end is forged into a yoke to take the eye of the valve rod. The pin has a ring and taper cotter pin.

365. Stephenson's Link Reversing Motion. A schematic layout of Robert Stephenson's link motion using two eccentrics, one for forward motion and the other for reverse motion. The valve rod ends in a die block which slides in the expansion links. The expansion link is lifted or lowered by the weighbar and lever. The neutral position is shown when neither eccentric will work. Economy of steam can be obtained after starting in full gear, by 'notching up', i.e. moving the lever nearer to neutral to restrict the valve rod movement, lessening the valve movement and opening.

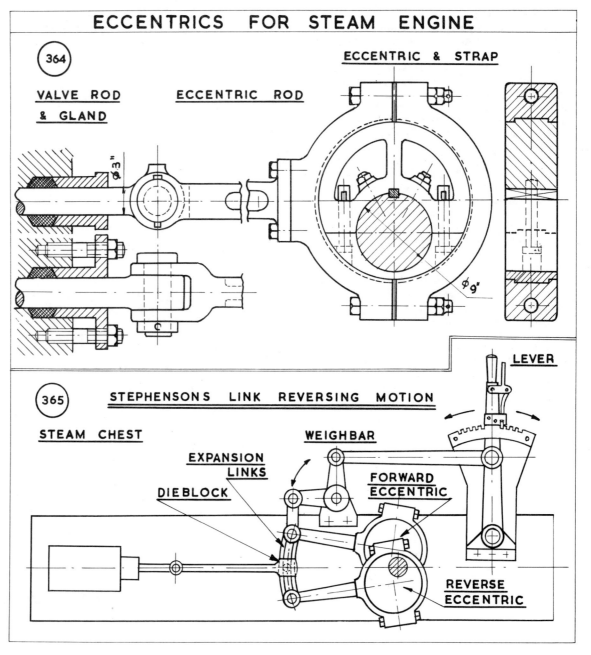

ECCENTRICS FOR STEAM ENGINE

364 ECCENTRIC & STRAP

VALVE ROD & GLAND   ECCENTRIC ROD

365 STEPHENSONS LINK REVERSING MOTION

LEVER

STEAM CHEST   WEIGHBAR

EXPANSION LINKS

DIEBLOCK

FORWARD ECCENTRIC

REVERSE ECCENTRIC

366. Internal Combustion Engines The two-stroke petrol engine draws in an explosive charge of petrol and air through a carburettor into the crankcase where it is compressed by the downward power stroke until the piston uncovers the transfer ports and allows the charge to travel to the cylinder where it is compressed by the rising piston and ignited by the spark.

At the end of the power stroke the exhaust gases pass out by the exhaust port. The cycle is completed in two strokes of the piston—once every revolution.

The lightweight piston—aluminium alloy—has a cast-iron ring to give a gas seal, the little end of the connecting rod has either a plain bush or needle rollers, the big end has a plain bearing, ball race or rollers. The crankshaft is usually built up, press fits or coned fits-nutted. The drive to the gear box is by sprocket and chain or spur gears. The ignition is by battery and coil, or flywheel magneto which generates the high tension spark to the spark plug. The timing of the spark is effected by a contact breaker with tungsten points opened by a cam on the crankshaft. The cooling of the cylinder is by air currents generated by the passage of the machine or by a shrouded fan driven by the engine itself. Lubrication is usually effected by adding oil to the petrol fuel, but variable oil pumps are now being fitted.

Two-stroke engines are made in working sizes from 50 c.c. to 350 c.c., popular range, and used for cycles, lawn-mowers, boat engines and industrial compressors and similar purposes. Racing two-stroke engines of 100 c.c. twin in line can now develop 25 b.h.p. at 7,500 r.p.m.

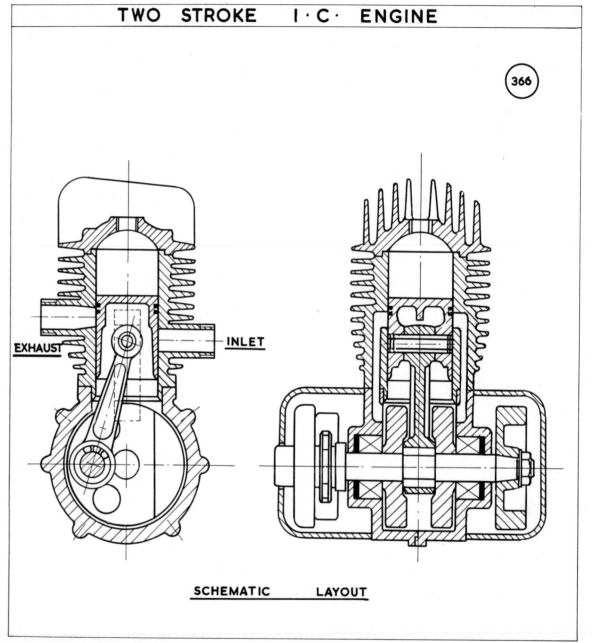

TWO STROKE I·C· ENGINE

366

EXHAUST

INLET

SCHEMATIC LAYOUT

367. Internal Combustion Engine Four-stroke. The four-stroke petrol internal combustion engine completes its cycle of induction, compression, power, exhaust in four strokes of the piston, i.e. two revolutions of the crankshaft. The control of the charge of petrol and air from the carburettor to the cylinder is by a cam-operated, mushroom-headed, poppet valve, and the exhaust gases leave on the opening of a similar valve. The valves may be duplicated giving four valves per cylinder to facilitate rapid induction and exhaust giving greater speed and power. Aluminium alloy pistons are generally used, sealed by cast-iron piston rings. The gudgeon pin is located by circlips, and the little end is a plain bush or needle rollers. The big end is a plain bearing or ball race or rollers. The valves are operated by push rods and tappets from a two to one reduction cam shaft from the main shaft. The ignition is by coil and battery, and a cam operated contact breaker gives the spark timing. The schematic diagram shows an air cooled, motor-cycle type, petrol engine.

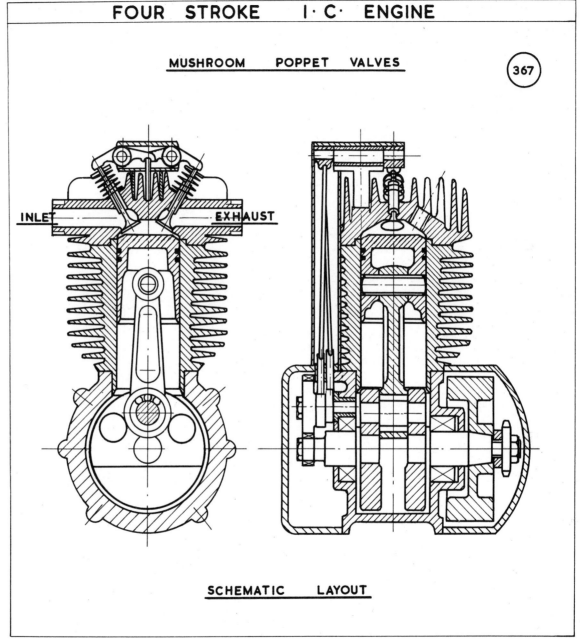

FOUR STROKE I·C· ENGINE

MUSHROOM POPPET VALVES

367

INLET EXHAUST

SCHEMATIC LAYOUT

368. I.C. Pistons and Connecting Rods. Two-stroke Type. Details of a piston used in a two-stroke engine are shown. The top of the piston is slightly domed, the skirt is cut away to suit the transfer ports, the gudgeon pin is located by a circlip.

The connecting rod is a forging with a bushed little end, and the big end is ready to receive the rollers. A connecting rod with a closed big end must be threaded on the crankpin during assembly of the built up crankshaft.

369. Four-stroke Piston and Connecting Rod. In some automobile engines 'over square' proportions have been adopted, i.e. the length of the stroke is less than the diameter of the piston, resulting in less piston movement for the same power. Larger valves are needed to facilitate breathing. Very short connecting rods can be used, large diameter plain bearings in the big ends, and since hollow cast or forged four- or six-throw crankshafts are used, split bearings are used. Units of this type have been used in four-cylinder car engines of 1,000 c.c. to 1,500 c.c.

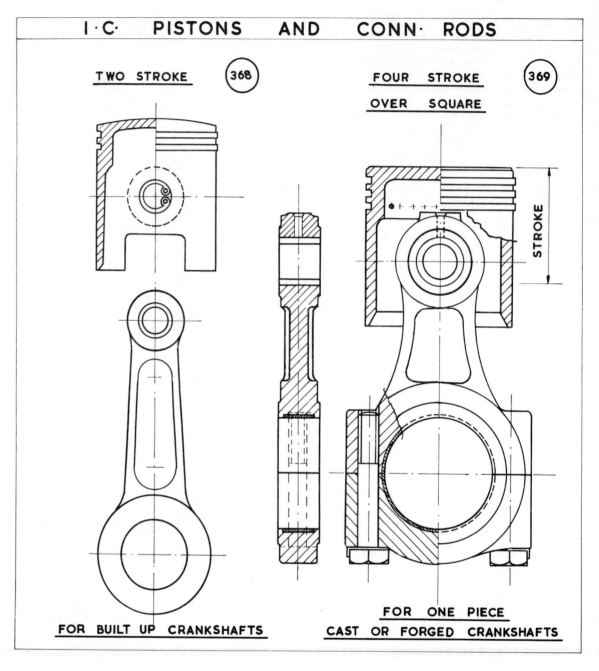

I·C· PISTONS AND CONN· RODS

TWO STROKE (368)

FOUR STROKE (369)
OVER SQUARE

STROKE

FOR BUILT UP CRANKSHAFTS

FOR ONE PIECE
CAST OR FORGED CRANKSHAFTS

370. I.C. Carburettor The carburettor is a means of obtaining a measured flow of petrol and air to the induction chamber. A diaphragm pump provides a steady flow of fuel to the float chamber and a needle valve and float ensure a steady level at the jets. A venturi tube with a reduced diameter giving an accelerated air flow over the jet picks up the fuel when the butterfly main valve is opened. For slow running and fast speeds extra jets meter the fuel to the engine's requirements. A choke valve, cutting off part of the air supply for starting purposes is also fitted.

371. Coil Ignition The ignition coil consists of a core of soft iron wire round which a primary winding of heavy copper wire is made. The secondary winding consists of many turns of very fine wire. When a low tension current of 12 volts is passed through the primary winding and broken, a high tension current is induced in the secondary winding of often 7,000 volts. This process can be used to produce a spark across the electrode of the plug in the cylinder head to ignite the petrol and air charge.

372. Contact Breaker Single cam operated contact breaker opens tungsten points to cause the igniting spark at the plug as the piston approaches T.D.C. A condenser across the points intensifies the spark.

373. Contact Breaker for four-cylinder engine.

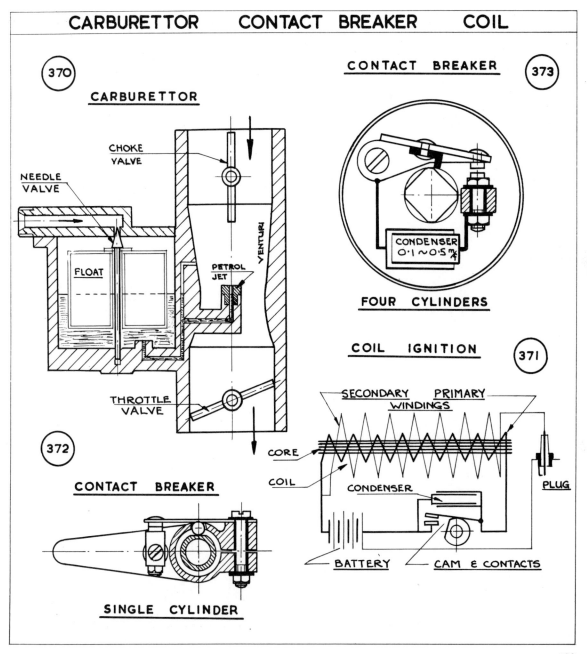

CARBURETTOR   CONTACT BREAKER   COIL

370   CARBURETTOR

CHOKE VALVE

NEEDLE VALVE

VENTURI

FLOAT

PETROL JET

THROTTLE VALVE

372   CONTACT BREAKER

SINGLE CYLINDER

CONTACT BREAKER   373

CONDENSER 0·1∼0·5 m/4

FOUR CYLINDERS

COIL IGNITION   371

SECONDARY   PRIMARY WINDINGS

CORE

COIL

CONDENSER

BATTERY   CAM & CONTACTS

PLUG

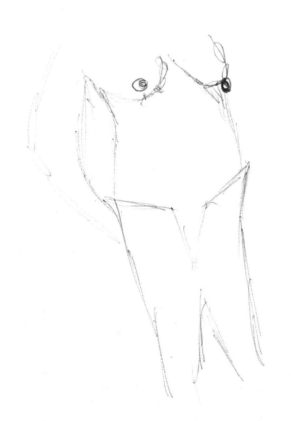

STANDARDS, LIMITS AND SYMBOLS

Standards

Conventions

Tolerances and Limits

Geometric Tolerances

Machining Symbols

Surface Texture Symbols

374. Standards and Conventions The British Standard 308:1964, covers fully the conventional methods of representing engineering components in drawing practice and is the recommended reference book for most examinations. It may be necessary to refer to the standard for fuller explanation on difficult points as many items are discussed in more detail than is possible in this volume.

A simple sans-serif block lettering and eight types of drawing lines are shown, and their special uses are shown in later drawings.

A simplified drawing layout, headings and parts list are shown. More involved layouts, suited to commercial drawing offices, will be found in the standard, but a simplified type is more suited to the time limits imposed by examination conditions.

A typical bracket section is shown, web not sectioned.

A simple example of machining symbol and tolerances; radial measurements showing diameters, using an abbreviation or the symbol, are also shown.

The conventional method of showing movement of an arm to a second position is shown in the drawing layout.

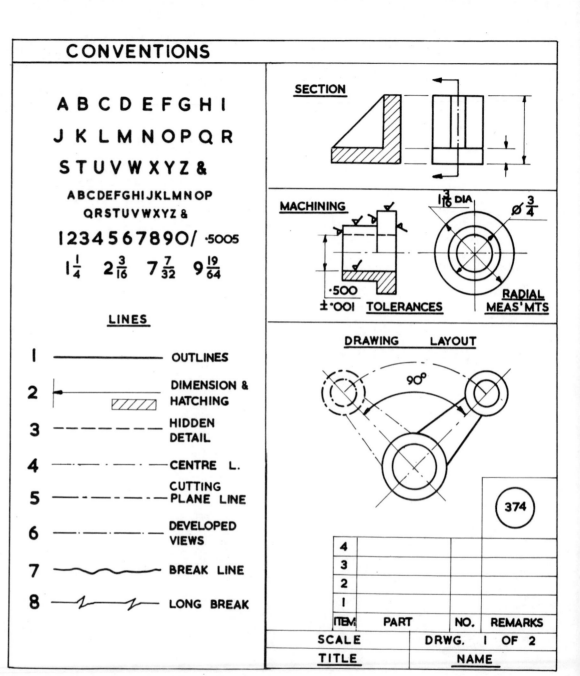

375. Conventions Counterbores and Countersinks.

A. The hole may be shown as a fully dimensioned diagram, or with a simplified diagram with printed instructions.

B. The hole may be shown as a fully dimensioned diagram or simplified with printed instructions.

C. In castings, the rough surface at the face of a drilled hole may require spot-facing with a suitable cutter to provide a flat surface at right angles to the bore enabling the nut or bolt head to seat without distortion.

D. Screwthread Conventional Form. A quick simplified method of drawing a bolt in position. Notice how the hatching crosses the thread lines only in the unoccupied portion of the internal thread.

E. Four methods of indicating a taper are shown in the standard, one only being shown. In this case, the taper is shown per inch length, with a tolerance.

F. Shaft and Hole. The hole is usually taken as the standard part, executed by drilling and reamering. A plus tolerance on the hole is usually allowed. The shaft is usually machined or ground to a minus tolerance to give the required running fit.

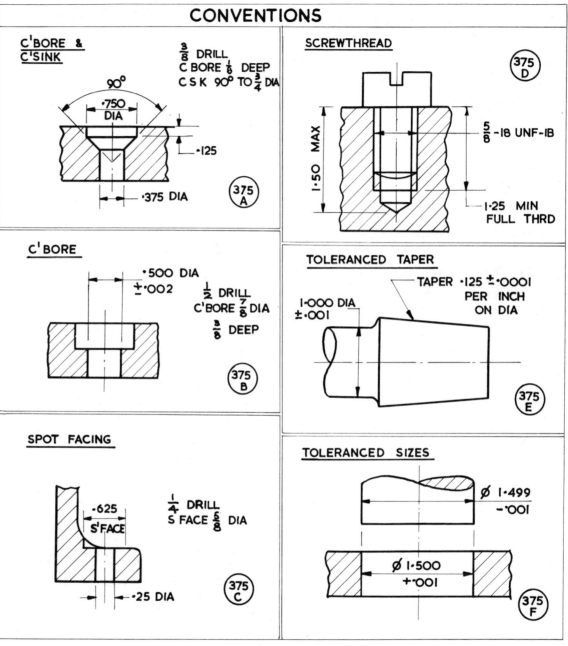

CONVENTIONS

C'BORE & C'SINK

$\frac{3}{8}$ DRILL
C BORE $\frac{1}{8}$ DEEP
C S K 90° TO $\frac{3}{4}$ DIA

90°
·750 DIA
·125
·375 DIA

375 A

C'BORE

·500 DIA ±·002

$\frac{1}{2}$ DRILL
C'BORE $\frac{7}{8}$ DIA
$\frac{3}{8}$ DEEP

375 B

SPOT FACING

·625
S'FACE
·25 DIA

$\frac{1}{4}$ DRILL
S FACE $\frac{5}{8}$ DIA

375 C

SCREWTHREAD

1·50 MAX
$\frac{5}{8}$ -18 UNF-1B
1·25 MIN FULL THRD

375 D

TOLERANCED TAPER

TAPER ·125 ±·0001 PER INCH ON DIA
1·000 DIA ±·001

375 E

TOLERANCED SIZES

Ø 1·499 -·001
Ø 1·500 +·001

375 F

193

376. Toleranced Dimensions Holes and Shafts. 'A toleranced dimension defines limits of size of a feature, and also has a bearing on the geometrical form of the feature'. B.S. 308.

Modern industry often requires that components are machined within specified tolerance zones of the nominal size, to allow interchangeability for assembly.

Holes and Shafts A shaft may be fitted into a hole to give:

(a) An interference fit, or press fit, in which the shaft is larger than the hole and requires pressure and/or heat for assembly.

(b) Key fits and drive fits, shaft slightly larger than hole, still giving rigid assembly.

(c) Location fits, shaft slightly less in diameter than the hole allowing withdrawal.

(d) Running fits, shaft smaller than hole allowing the shaft to revolve.

The 'hole' basis is usually adopted, in which the hole is the nominal size carrying a plus tolerance, since holes are easiest produced by drilling and reamering, and the shaft being machined to a plus or minus tolerance depending on the type of fit.

An average running fit and a precision running fit are shown in the lower diagram; the exaggerated hatched portion should help in explaining the method. The upper diagram shows a range of the four fits indicated on a graph tabled in half-thousandths of an inch, to a greatly enlarged scale. A fuller explanation and range is given in B.S. 1916:L953.

In dimensioning a drawing, the limits may be shown by (a) showing both limits, (b) size and limit in one direction, (c) size with tolerance above and below. See tables in Appendix.

377. Positional Tolerances Lengths, such as those shown in the diagram should be measured from a datum line to avoid multiple errors occurring.

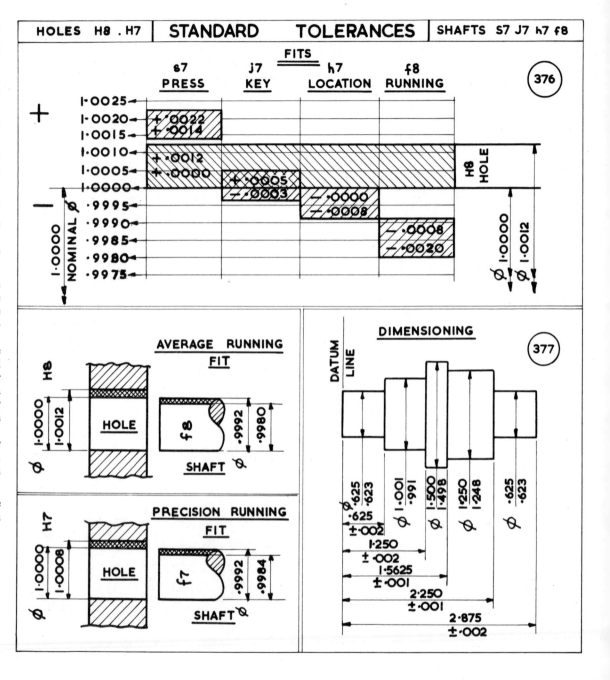

194

378. Geometrical Tolerances Geometrical Tolerances of straightness, flatness, parallelism, squareness, angularity, concentricity, symmetry and position, are required to ensure that component parts are interchangeable for assembly and function.

A. Straightness Str. Tol. The diagram shows the tolerance zone of 0·003″ applied to a cylindrical unit. The zone may be stated at the circumference or at the axis.

B. Flatness Flat. Tol. A tolerance zone may refer to the flatness of a certain surface, or to the non-tolerance of concavity of the surface.

C. Parallelism Par. Tol. The surface is considered in relation to another stated surface, the datum face, from which the depth measurement and the tolerance are calculated.
    Parallelism may also be required to toleranced limits between a datum face and a hole.
    Parallelism may also be required to toleranced limits between a datum axis of a stated hole and another hole.

D. Squareness Sq. Tol. Squareness tolerance between two faces at right angles to each other within the specified limit.
    Squareness of a cylinder to a datum face to a tolerance.
    Squareness of a surface at right angles to a datum shaft.
    Squareness of a hole to a datum hole to toleranced limits.

E. Angular Tolerance Ang. Tol. The tolerated limit of error of a face at a specified angle to a datum face.

B.S. 308.

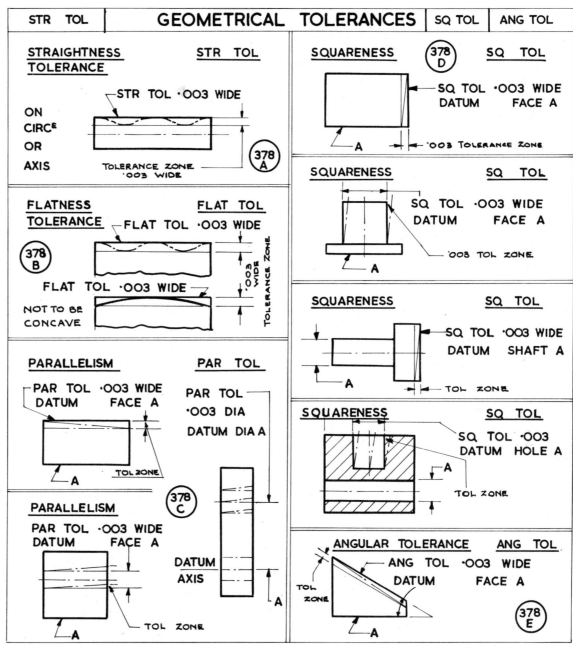

379. Geometrical Tolerances

A. Concentricity Conc. Tol. Concentricity of a cylinder to a datum cylinder within a stated tolerance zone.

Concentricity tolerance of related cylinders to a specified zone.

Concentricity of two cylinders A and B datum, and a third cylinder within a tolerance zone.

B. Symmetry Sym. Tol. Symmetrical tolerance of a zone between a datum width A and a slot.

Symmetry Toleranced error within a zone, of two widths.

Symmetry Toleranced position of a hole in relation to two datum widths A and B, within a zone.

B.S. 308.

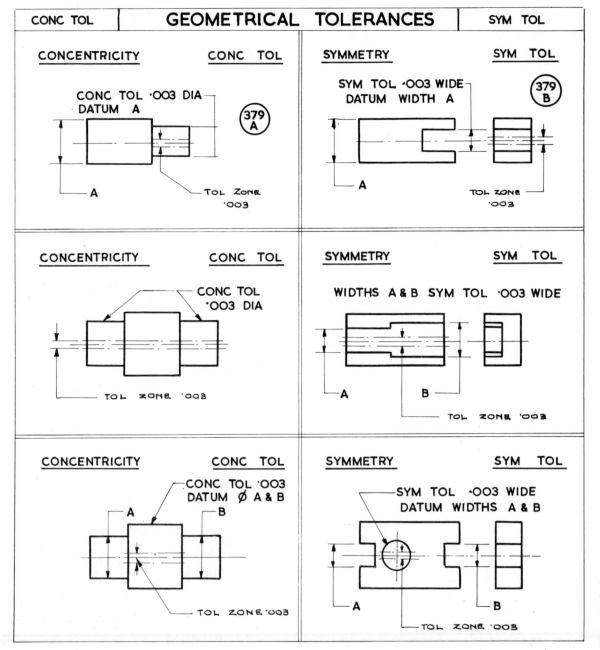

380. Geometrical Tolerances

A. Position Tolerance Posn. Tol. Positional tolerance of the centre of a hole from two lines. TP indicates True Position.

  Position Three holes positioned within a stated tolerance of the true position.

  Position Holes positioned from a datum hole A and a face B on a pitch circle diameter and an angularity true position.

B. Maximum Metal Condition (MMC) and **Minimum Metal Condition** The diagrams show the two situations of a standard hole, and two cylinders, machined, one to the lower limit, and the other to the upper tolerance limit.

C. Combined Tolerances The diagram shows a component in which concentricity and squareness tolerances are stated in relation to a datum face D, within the maximum metal condition (MMC).

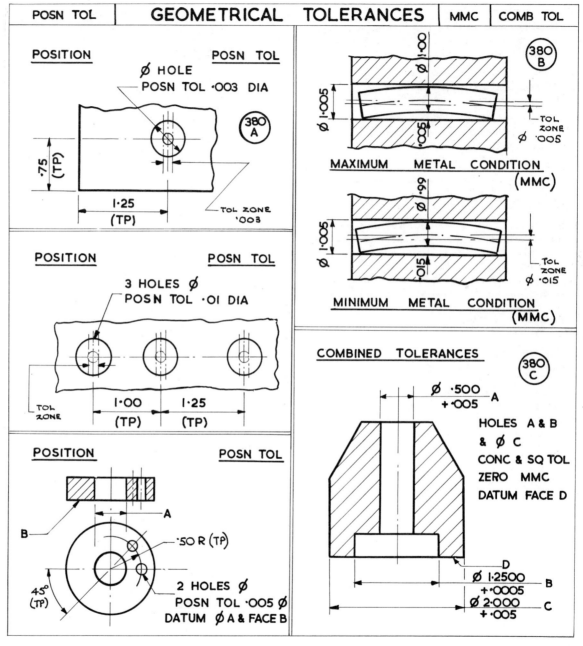

381. Machining and Surface Texture Symbols

A. Machining Symbol Where a surface is to be machined to normal limits, a machining symbol may be shown as given to indicate such surfaces.

B. General Use of the Symbol The diagram shows the use of the symbol. A note on the drawing may also be used to indicate the limits of the surface finish.

C. Special Instructions Where the roughness tolerance is to be stated in micro-inches, the symbol should include the index number. Maximum and minimum values can be stated in the symbol as shown.

D. Special Process If a special process such as lapping or honing is desired to give a finish to certain limits, the symbol should bear the instruction and the index number.

E. Direction of Lay Where the direction of the application of the finish is specially required, the direction of the lay, either lateral or circular should be stated and the index number included with the symbol.

The index numbers used in B.S. 1134 indicate surface roughness measured in micro-inches; one micro-inch being 0·000001″.

Suggested index numbers are; 1, 2, 4, 8, 16, 32, 63, 125, 500, 1000.

F. Surfaces may be finished by turning, shaping and milling to ordinary workshop limits down to 0·0005″; finer limits by grinding, lapping, honing and scraping.

Special surface treatments such as plating, anodising and chemical process, should be explained in a suitable note in the drawing.

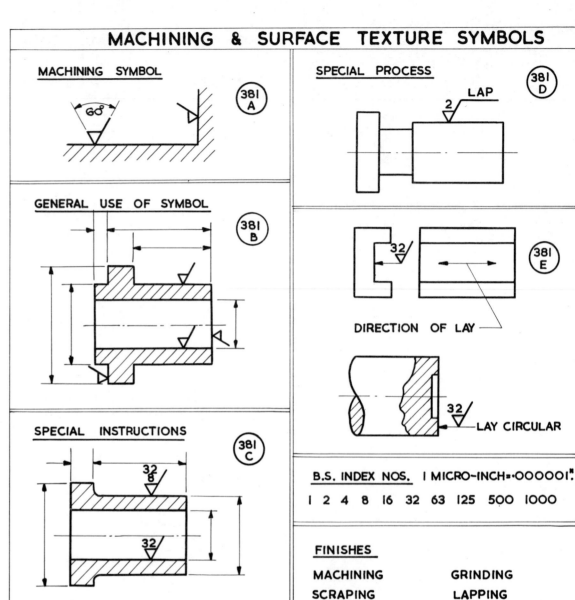

198

MACHINE DRAWING PROBLEMS

Pump Parts

Cover Plate

Vane Pump Parts

Bearing Bracket

Face Valve Parts

Gearbox End Plate

Camshaft Pump

Worm Reduction Gear Parts

Screw Pump Parts

Cylinder Head

Overhead Swivel Pulley

Gear Box

Swashplate Pump

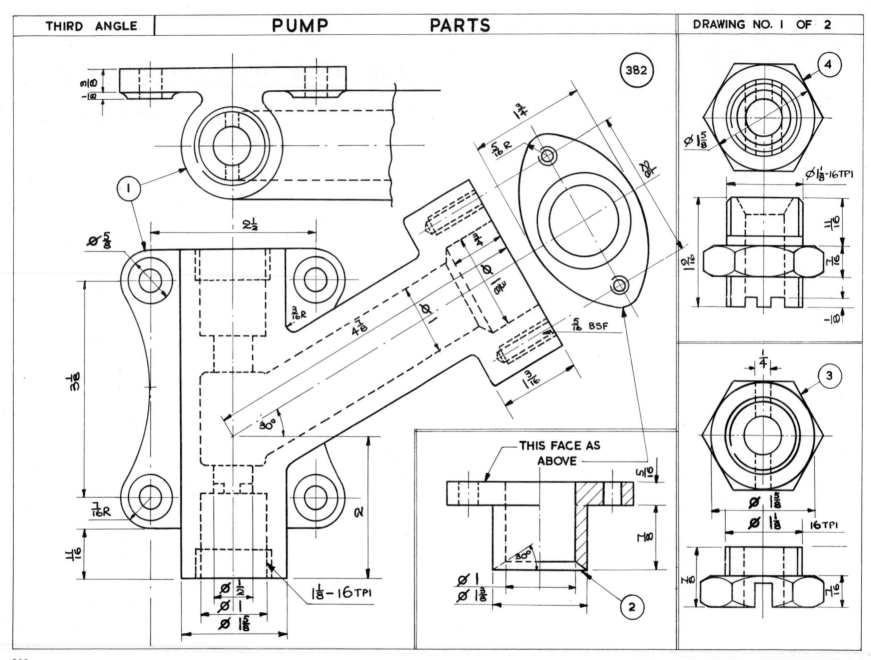

THIS FACE AS ABOVE

$1\frac{1}{8}$–16TPI

$\emptyset 1\frac{1}{8}$-16TPI

$\emptyset 1\frac{5}{8}$

16TPI

$\frac{5}{16}$ BSF

30°

MACHINE DRAWING PROBLEMS

The following drawings are set as examination type questions. Some of the data will require a search to be made to test ability both in reasoning and in visualising the parts and assembly of the unit.

Setsquare angles may be inferred, details and sizes not shown are left to discretion. Most of the problems show parts from which assembly views are required, and reference to solutions and isometric drawings shown later should not be made until some attempt at a solution has been attempted, to simulate examination conditions. A number of the objects set are complete working units such as pumps, and would form suitable projects for construction in the workshop if time permits, bringing the work to an ideal conclusion.

382 and 383. Pump Parts The parts of a water pump with an inclined ram are given. The valve seat screws into the lower end of the body casting, a plastic or metal ball valve seating upon it. The upper ball valve is retained by the union. The ram is sealed by the hemp ring and gland fastened by two bolts.

Draw:

(a) In First Angle Projection, a sectional front elevation with the parts assembled, ram at lowest position, f.s.
(b) Project a full plan from the required elevation.
(c) Project a full end elevation to the left of the front elevation.

Complete the drawing by adding a list of parts, title, border, and six dominant dimensions. A part solution is given later, see 408.

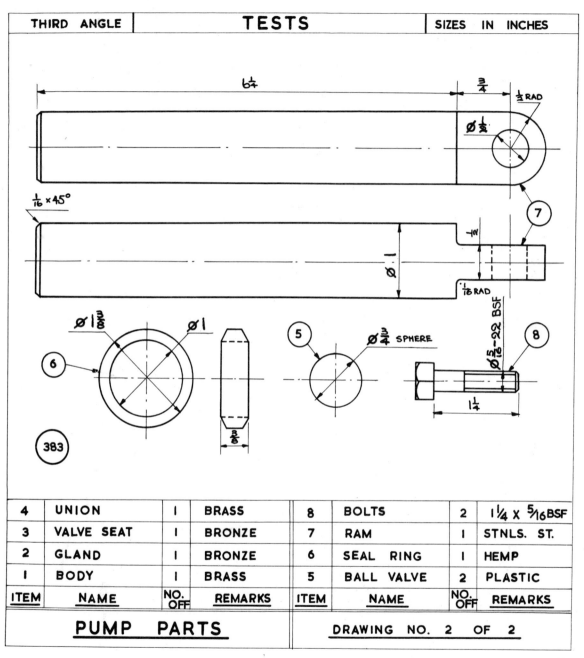

4	UNION	I	BRASS	8	BOLTS	2	1¼ X ⁵⁄₁₆ BSF
3	VALVE SEAT	I	BRONZE	7	RAM	I	STNLS. ST.
2	GLAND	I	BRONZE	6	SEAL RING	I	HEMP
I	BODY	I	BRASS	5	BALL VALVE	2	PLASTIC
ITEM	NAME	NO. OFF	REMARKS	ITEM	NAME	NO. OFF	REMARKS

PUMP PARTS	DRAWING NO. 2 OF 2

384. Cover Plate A cover plate with bearings for three shafts is shown in Third Angle projection, sizes in millimetres.

Draw:

(a) In First Angle, f.s., the given plan as the new front elevation.

(b) Project an end elevation to the left of the new front elevation.

(c) Project a plan in accordance with the two above views.

(d) Project an Auxiliary Plan from the new front elevation normal to a line drawn through the two centres of the larger shafts, sectional.

Dimension the front and end elevations, and the plan. Add titles and name. Symbol machined surfaces. Add tolerances for average fit shaft holes.

(e) Make a full size isometric drawing, natural scale, of the cover plate, with the smallest shaft hole nearest the viewer.

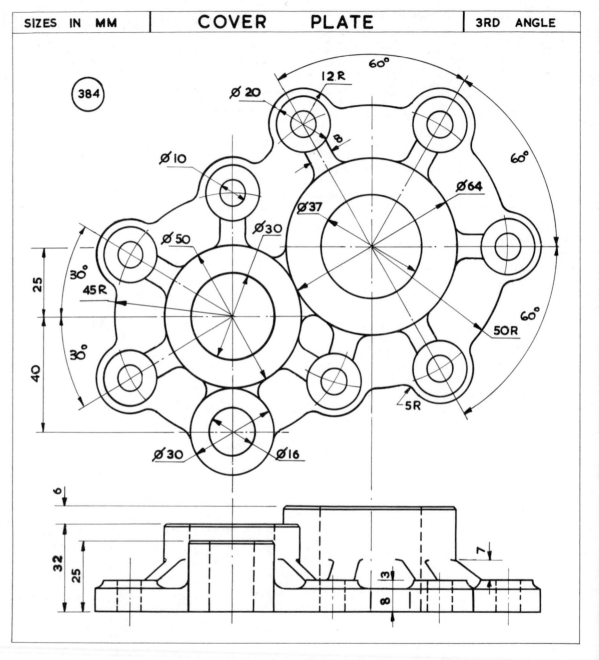

| SIZES IN MM | COVER PLATE | 3RD ANGLE |

202

385 and 386. Vane Pump Parts The parts of a rotary vane pump are given. The sliding blades of the rotor are kept in contact with the housing by centrifugal force. The rotor turns in a clockwise direction, fluid entering the top orifice and being pumped out of the lower opening. A sealing ring and gland are held in place with two bolts. A cover plate closes the front of the housing sealed by a $\frac{1}{64}''$ gasket.

Draw:

(a) F.S. using First Angle Projection, a full elevation of the pump with front cover removed but otherwise assembled, and looking at the end of the rotor.

(b) Project a full elevation to the right of the front elevation.

(c) Project a plan, sectioned on the centre line of the rotor shaft.

(d) Draw a union and cone joint to suit $\frac{3}{8}''$ o.d. tube the pump connection.

Complete the drawing by adding a list of parts, titles. Show six overall sizes.

See 409 for an assembly drawing.

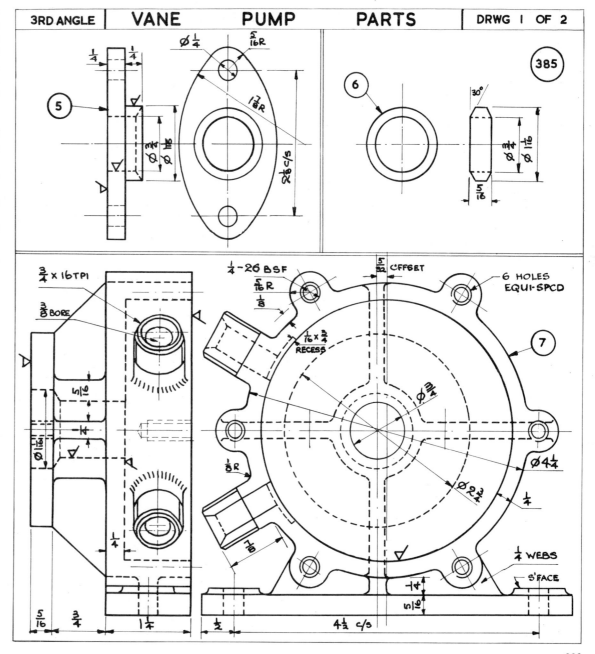

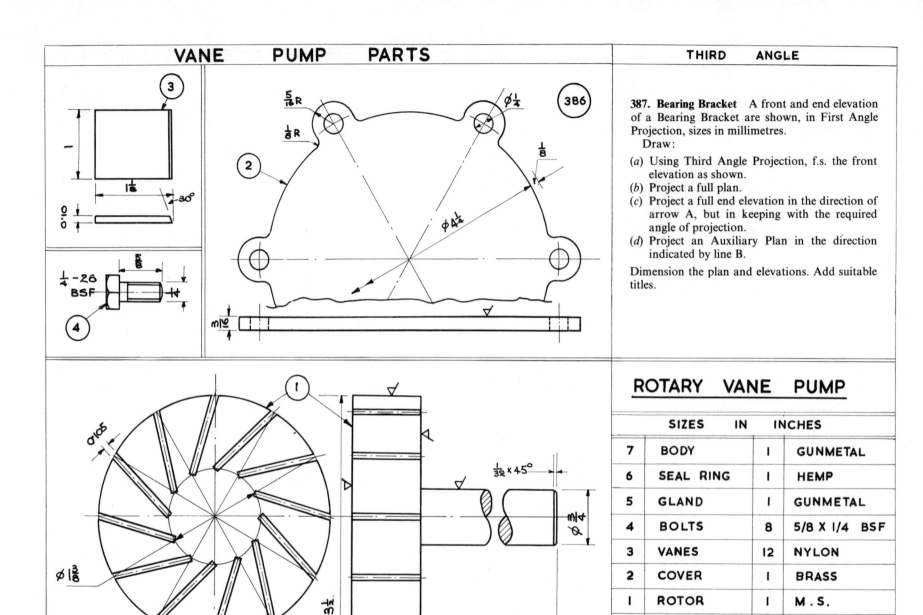

387. Bearing Bracket A front and end elevation of a Bearing Bracket are shown, in First Angle Projection, sizes in millimetres.

Draw:

(a) Using Third Angle Projection, f.s. the front elevation as shown.

(b) Project a full plan.

(c) Project a full end elevation in the direction of arrow A, but in keeping with the required angle of projection.

(d) Project an Auxiliary Plan in the direction indicated by line B.

Dimension the plan and elevations. Add suitable titles.

ROTARY   VANE   PUMP

		SIZES	IN	INCHES
7	BODY		I	GUNMETAL
6	SEAL RING		I	HEMP
5	GLAND		I	GUNMETAL
4	BOLTS		8	5/8 X 1/4 BSF
3	VANES		12	NYLON
2	COVER		I	BRASS
I	ROTOR		I	M.S.
ITEM	NAME		NO.	REMARKS

DRAWING   NO. 2 OF 2

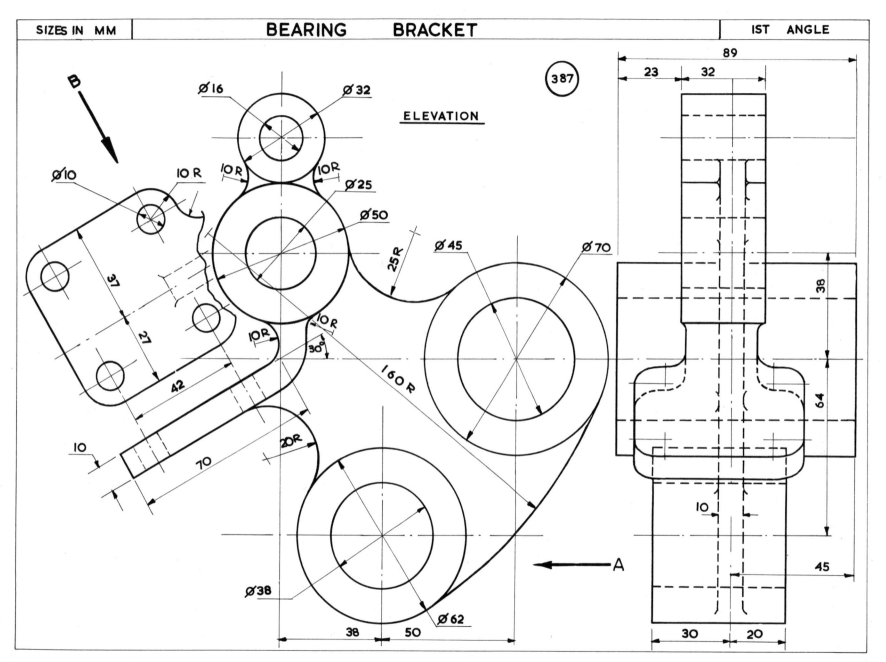

ELEVATION

387

ØI6
Ø32
ØIO
IOR
IOR
IOR
Ø25
Ø50
37
27
42
IOR
IOR
30°
I60R
IO
70
20R
Ø45
Ø70
25R
Ø38
Ø62
38
50

89
23
32
38
64
45
IO
30
20

← A
B

205

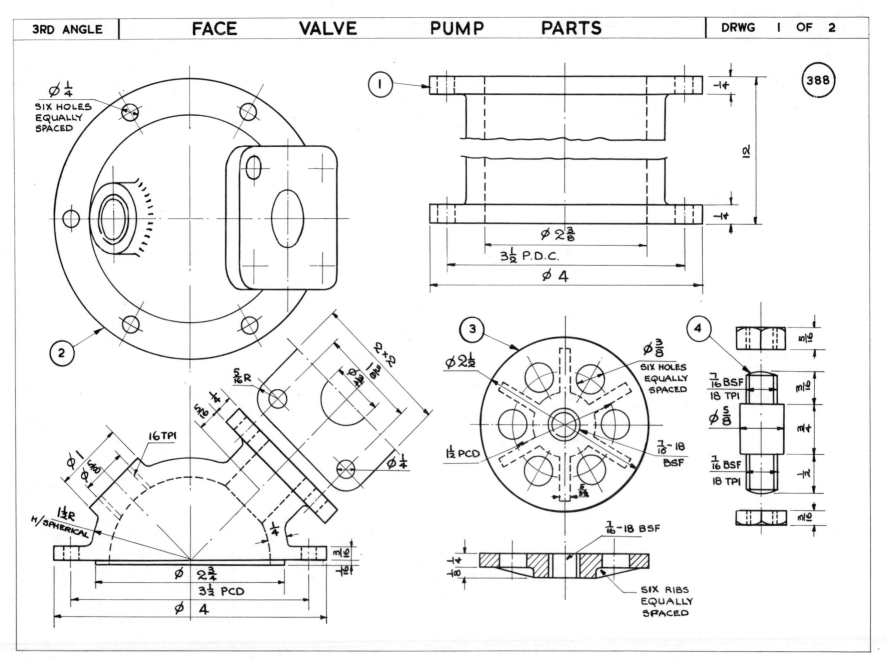

388 and 389. Face Valve Parts Parts of the pump are shown. The pin screws into the valve plate; the diaphragm, washers and spring fit on the pin and are locked in place by the nut. The locknut secures the valve plate. The valve assembly drops into the top of the cylinder, and is held by the cylinder head.

The safety-release valve screws, as an assembly, into the cylinder head.

Draw:

(a) F.S. First Angle Projection, a front sectional elevation of the parts assembled taken on a vertical centre line of the cylinder.

(b) Project a full plan from the elevation.

(c) Project an end elevation to the left of the sectional elevation.

The cylinder height may be reduced by break lines to save drawing space.

A part assembly diagram is given on another page, Third Angle, see 410.

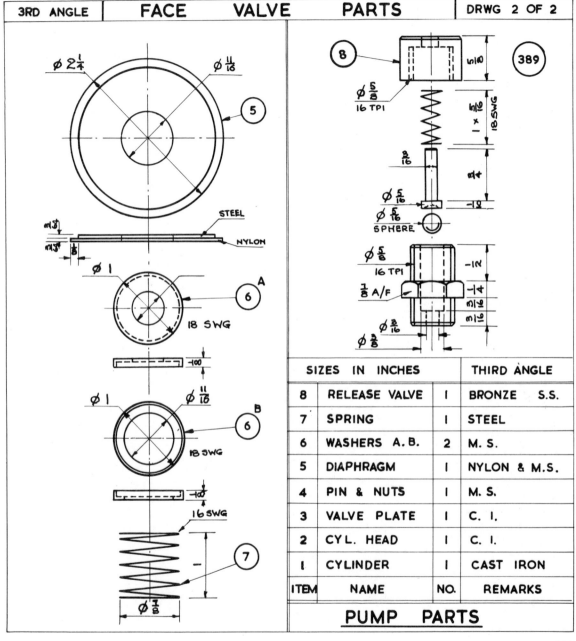

SIZES IN INCHES			THIRD ANGLE
8	RELEASE VALVE	1	BRONZE S.S.
7	SPRING	1	STEEL
6	WASHERS A. B.	2	M. S.
5	DIAPHRAGM	1	NYLON & M.S.
4	PIN & NUTS	1	M. S.
3	VALVE PLATE	1	C. I.
2	CYL. HEAD	1	C. I.
1	CYLINDER	1	CAST IRON
ITEM	NAME	NO.	REMARKS

PUMP PARTS

207

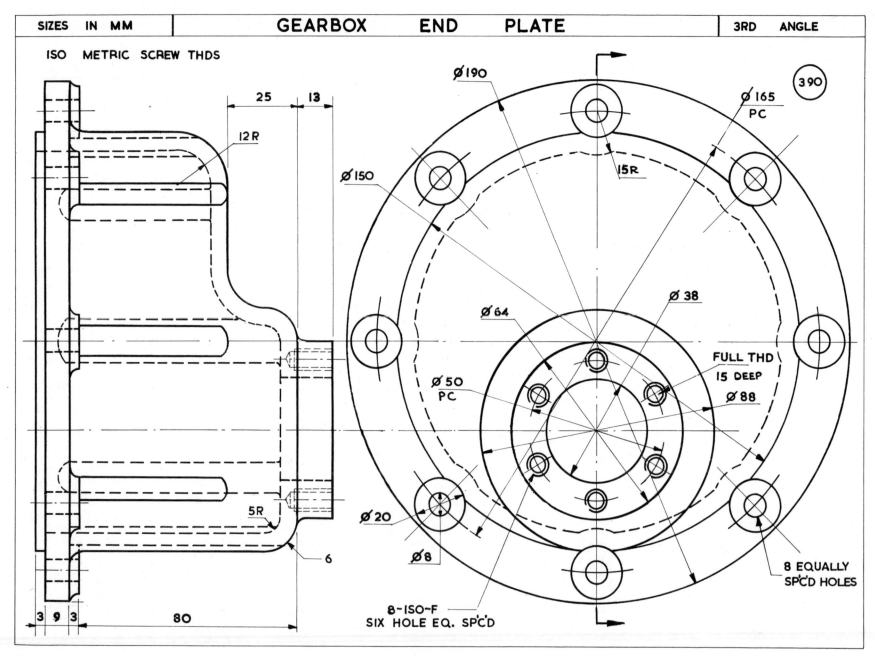

SIZES IN MM GEARBOX END PLATE 3RD ANGLE

ISO METRIC SCREW THDS

390

25 13

12R

Ø190

Ø165 PC

Ø150

15R

Ø 38

Ø 64

FULL THD 15 DEEP

Ø 88

Ø 50 PC

5R

Ø 20

Ø8

6

8 EQUALLY SP'C'D HOLES

8-ISO-F SIX HOLE EQ. SP'C'D

3 9 3 80

390. Gear Box End Plate Third Angle Projections of a gearbox end plate are shown.

Draw:

(a) F.S. First Angle Projection, a front elevation looking into the recessed interior of the plate.

(b) Project a sectional end elevation to the right of the front elevation.

(c) Project a plan. Dimension the diagrams fully. Add titles. Indicate surfaces requiring machining.

391, 392 and 393. Camshaft Pump Three sheets of diagrams showing parts of the camshaft pump are given. The body casting is bored to take ball races which receive the camshaft. This is retained in place by the cover plate. The two rams are held in contact with the cam faces by springs. The two inlet and outlet valves are stainless steel balls.

Draw:

(a) F.S., First Angle Projection, a sectional front elevation taken parallel to the axis of the camshaft.

(b) Project a full plan.

(c) Project a sectional end elevation to right of the front elevation, and on the line through the centre line of the first cam.

(d) Project a full end elevation to the left of the front elevation. Add eight major overall dimensions. Title.

(e) Draw 3 × f.s. a diagram showing how the contour of the cam is obtained. Uniform acceleration and retardation, lift $\frac{3}{8}''$.

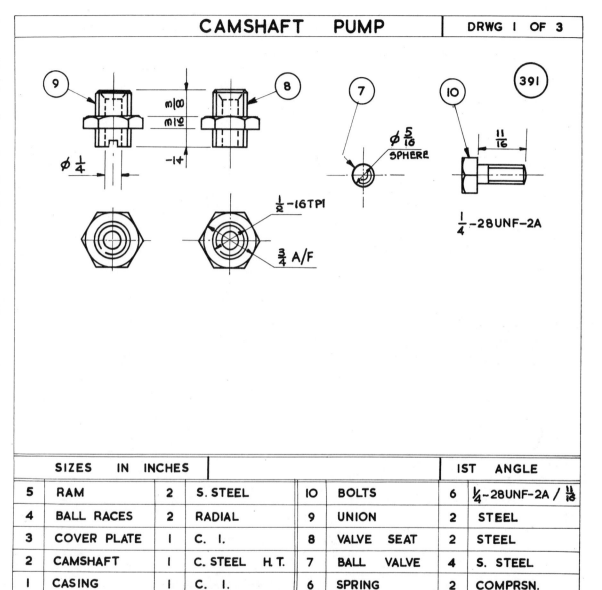

| CAMSHAFT PUMP | | | | DRWG I OF 3 |

	SIZES IN INCHES						1ST ANGLE
5	RAM	2	S. STEEL	10	BOLTS	6	$\frac{1}{4}$-28UNF-2A / $\frac{11}{16}$
4	BALL RACES	2	RADIAL	9	UNION	2	STEEL
3	COVER PLATE	I	C. I.	8	VALVE SEAT	2	STEEL
2	CAMSHAFT	I	C. STEEL H. T.	7	BALL VALVE	4	S. STEEL
I	CASING	I	C. I.	6	SPRING	2	COMPRSN.
ITEM	NAME	NO	REMARKS	ITEM	NAME	NO	REMARKS
	CAMSHAFT PUMP PARTS						

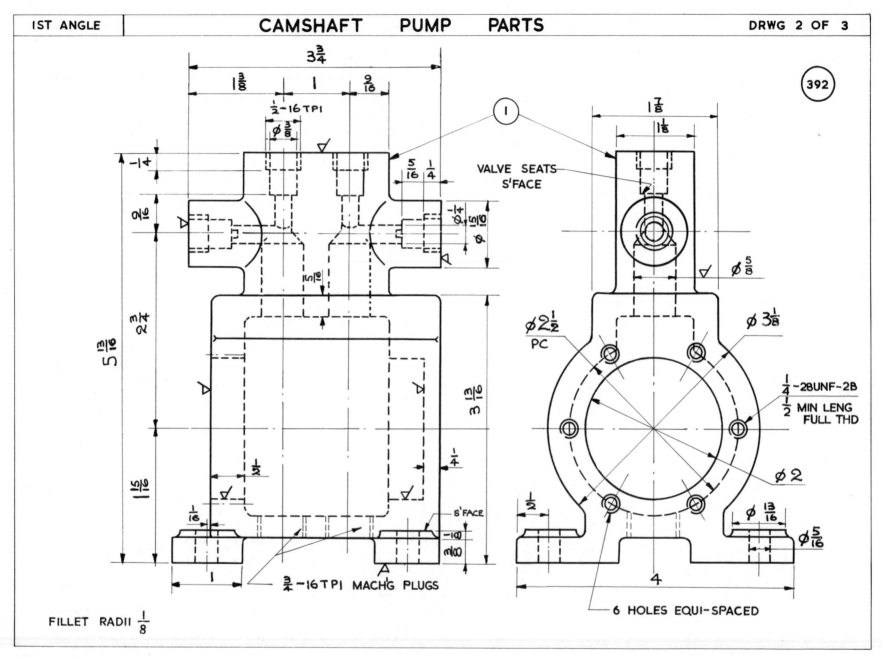

VALVE SEATS
S'FACE

¼-28UNF-2B
½ MIN LENG
FULL THD

6 HOLES EQUI-SPACED

¾-16TPI MACH'G PLUGS

S'FACE

FILLET RADII ⅛

210

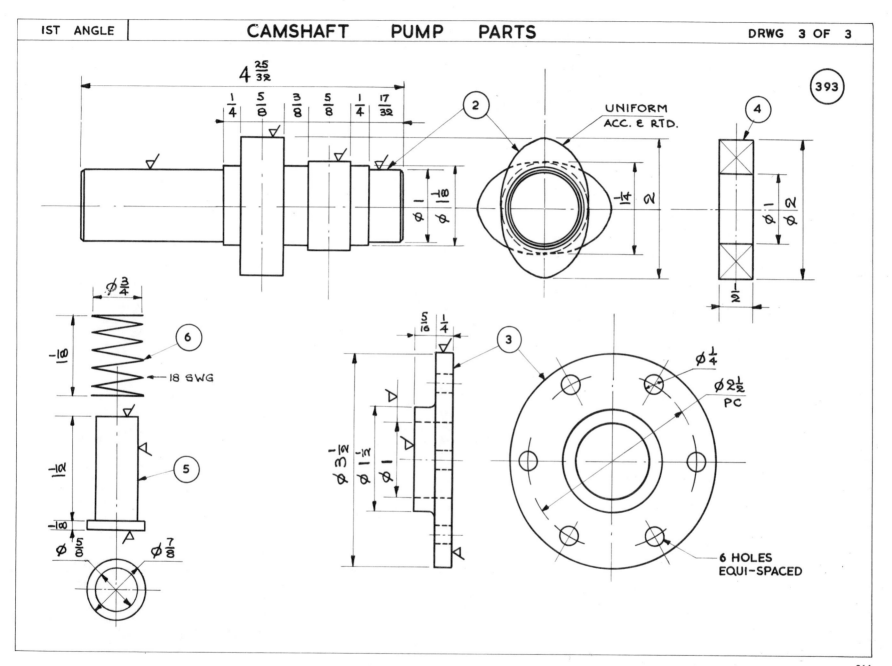

UNIFORM
ACC. & RTD.

393

18 SWG

6 HOLES
EQUI-SPACED

394, 395, 396 and 397. Worm Reduction Gear Parts The parts of a worm reduction gear are given. The wormwheel and shaft run in bushes set in the gear housing. The worm runs in bushes set in the casing and bearing support. A thrust race fits on the shorter end of the worm shaft. The wormwheel has twelve teeth. The worm tooth section is a rack section.

Draw:

(a) F.S. First Angle, a sectional elevation taken on the centreline of the worm with all the parts assembled. The worm may be drawn by simple representational straight lines, full helical curves need not be shown. Draw only three meshing teeth of the wormwheel, compass curves allowed.

(b) Project a full plan, from the elevation.

(c) Project a full end elevation which shows the bearing support.

Complete the drawing by adding a list of parts and titles. Add ten important dimensions.

Assembly diagrams are given on another page, Third Angle, see 413.

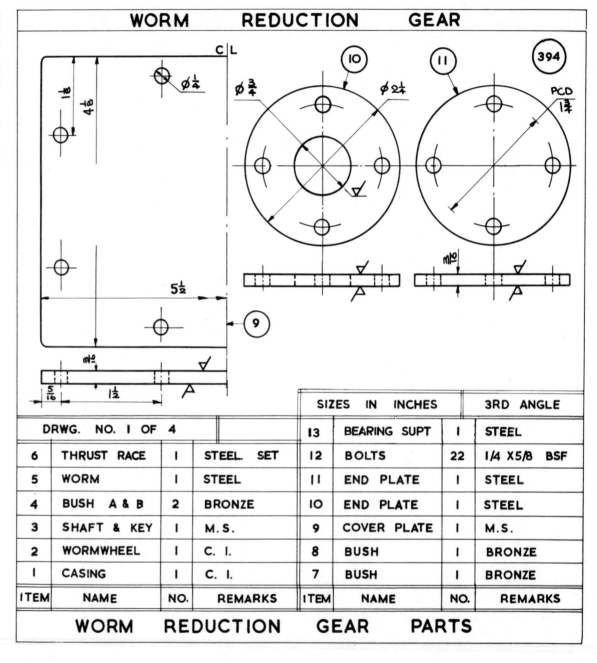

WORM   REDUCTION   GEAR

SIZES IN INCHES			3RD ANGLE		
DRWG. NO. I OF 4			13	BEARING SUPT	I	STEEL
6	THRUST RACE	I	STEEL. SET	12	BOLTS	22	1/4 X5/8 BSF
5	WORM	I	STEEL	II	END PLATE	I	STEEL
4	BUSH A & B	2	BRONZE	IO	END PLATE	I	STEEL
3	SHAFT & KEY	I	M.S.	9	COVER PLATE	I	M.S.
2	WORMWHEEL	I	C. I.	8	BUSH	I	BRONZE
I	CASING	I	C. I.	7	BUSH	I	BRONZE
ITEM	NAME	NO.	REMARKS	ITEM	NAME	NO.	REMARKS

WORM   REDUCTION   GEAR   PARTS

FILLET RADII $\frac{1}{8}$

①

395

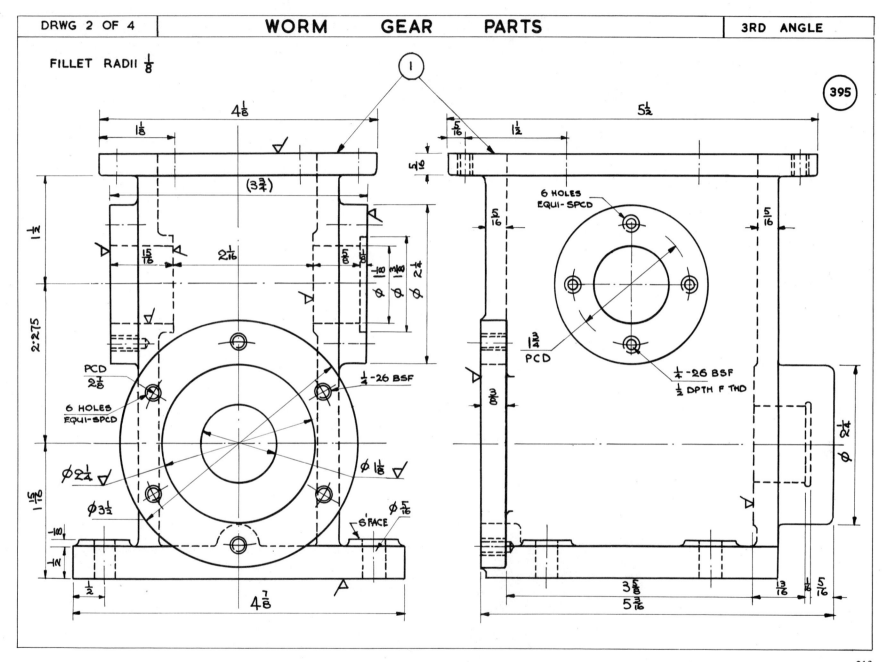

213

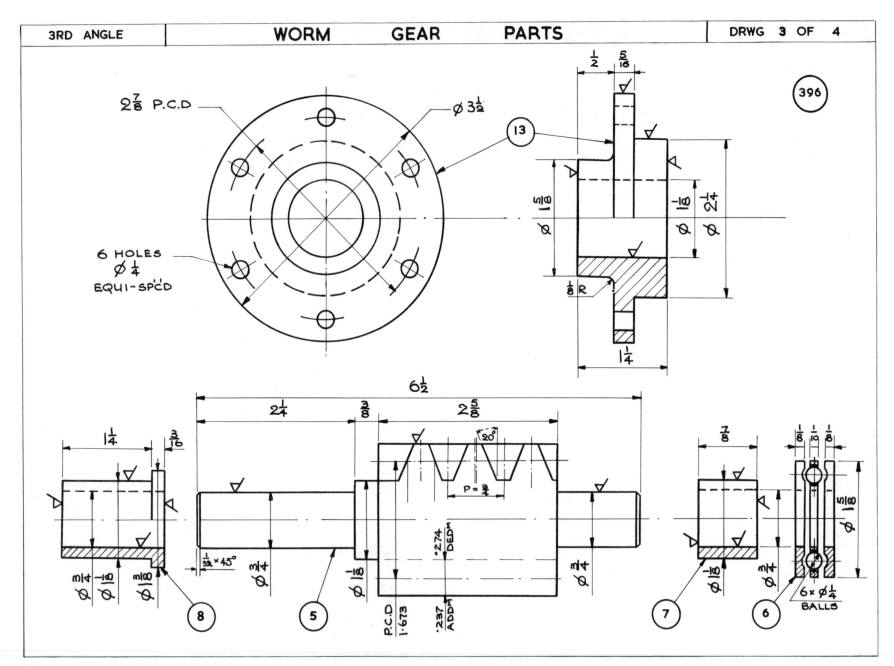

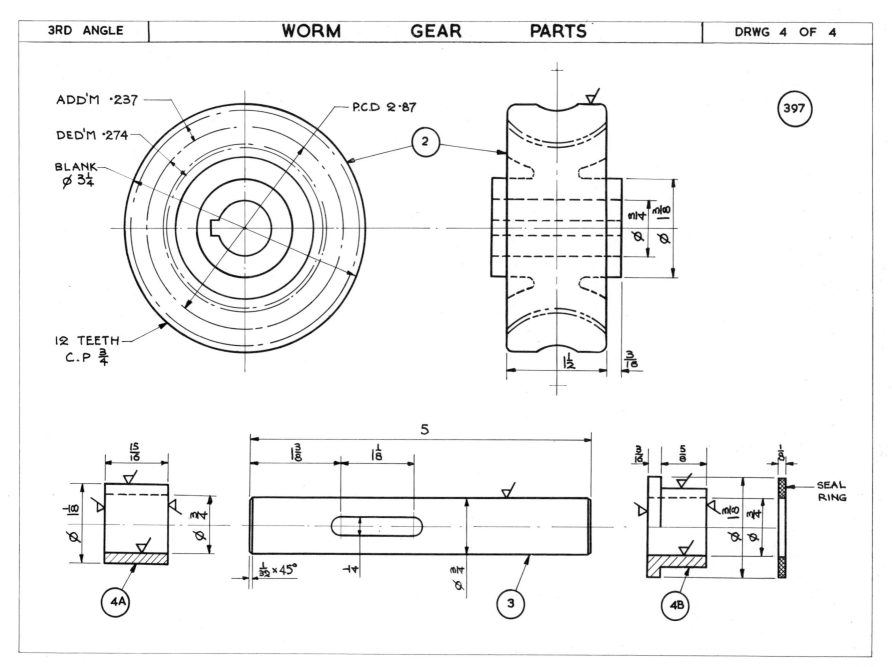

ADD'M ·237

DED'M ·274

BLANK Ø 3¼

P.C.D 2·87

2

12 TEETH C.P ¾

Ø ¾ Ø ⅜

1½

3/16

5

1⅜

1⅛

15/16

⅛

Ø ⅛ Ø ¾

4A

1/32 × 45°

¼

Ø ¾

3

3/16 ⅝ ⅛

Ø ⅜ Ø ¾

SEAL RING

4B

215

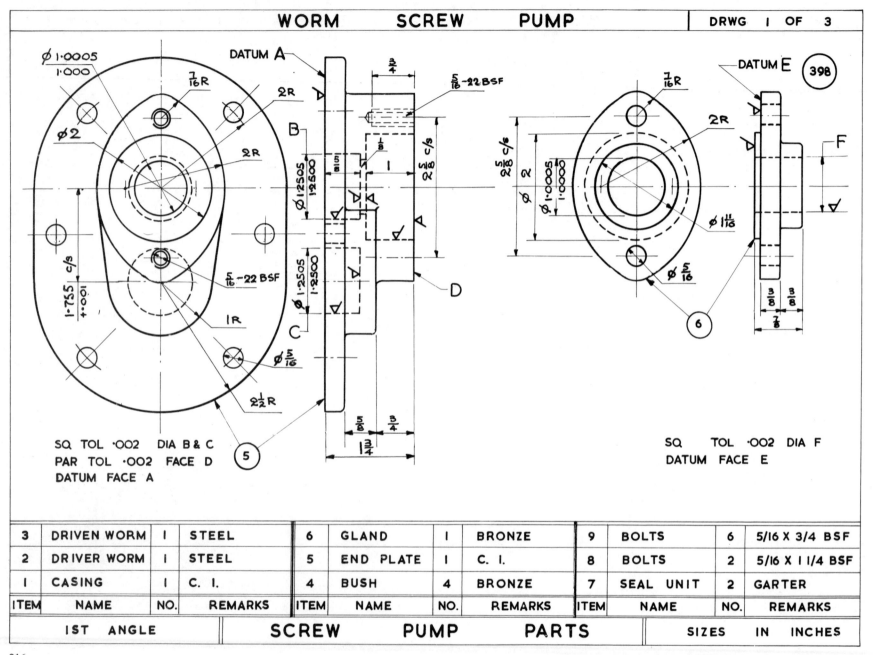

DATUM A

DATUM E   398

Ø 1·0005
1·000

7/16 R

2R

Ø 2

2R

2R

3/4

5/16 - 22 BSF

B

1/8

Ø 1·2505
1·2500

5/8

2 5/8 c/s

Ø 1·2505
1·2500

D

C

1R

5/16 - 22 BSF

1·755 c/s
+ ·001

2 1/2 R

Ø 5/16

5

SQ TOL ·002 DIA B & C
PAR TOL ·002 FACE D
DATUM FACE A

5/8   3/4

1 3/4

7/16 R

2R

2 5/8 c/s

Ø 1·0005
1·000

Ø 2

Ø 1 11/16

Ø 5/16

6

F

3/8   3/8

7/8

SQ TOL ·002 DIA F
DATUM FACE E

3	DRIVEN WORM	1	STEEL	6	GLAND	1	BRONZE	9	BOLTS	6	5/16 X 3/4 BSF
2	DRIVER WORM	1	STEEL	5	END PLATE	1	C. I.	8	BOLTS	2	5/16 X 1 1/4 BSF
1	CASING	1	C. I.	4	BUSH	4	BRONZE	7	SEAL UNIT	2	GARTER
ITEM	NAME	NO.	REMARKS	ITEM	NAME	NO.	REMARKS	ITEM	NAME	NO.	REMARKS
1ST ANGLE			SCREW PUMP PARTS					SIZES IN INCHES			

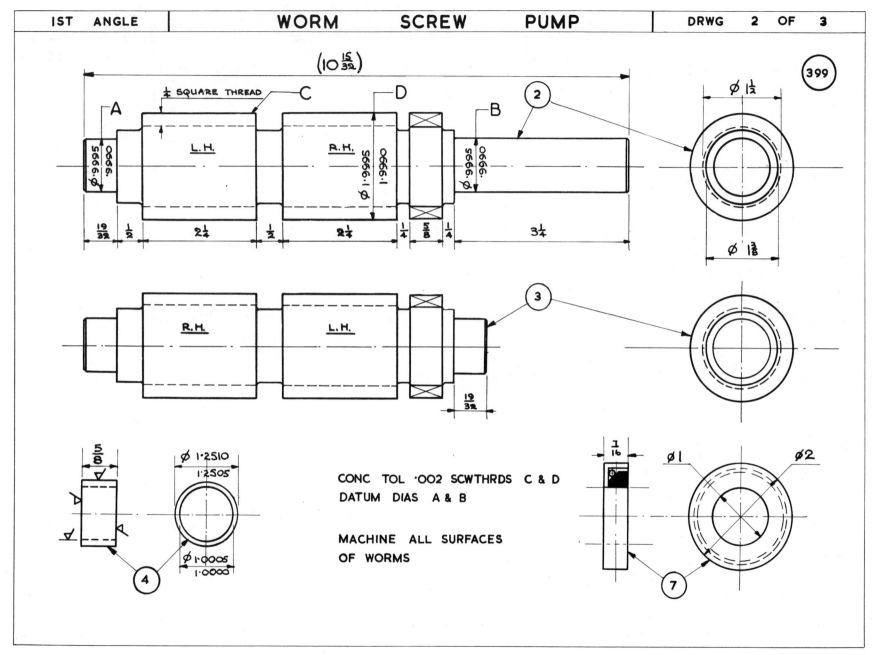

(10 15/32)

¼ SQUARE THREAD

A   C   D   B   2

.9990
.9995 Ø

L.H.   R.H.

399

Ø 1½

Ø 1⅞

19/32 ½ 2¼ ½ 2¼ ¼ 5/8 ¼ 3¼

R.H.   L.H.   3

19/32

5/8

Ø 1·2510
1·2505

Ø 1·0005
1·0000

4

CONC TOL ·OO2 SCWTHRDS C & D
DATUM DIAS A & B

MACHINE ALL SURFACES
OF WORMS

1/16

Ø1   Ø2

7

15

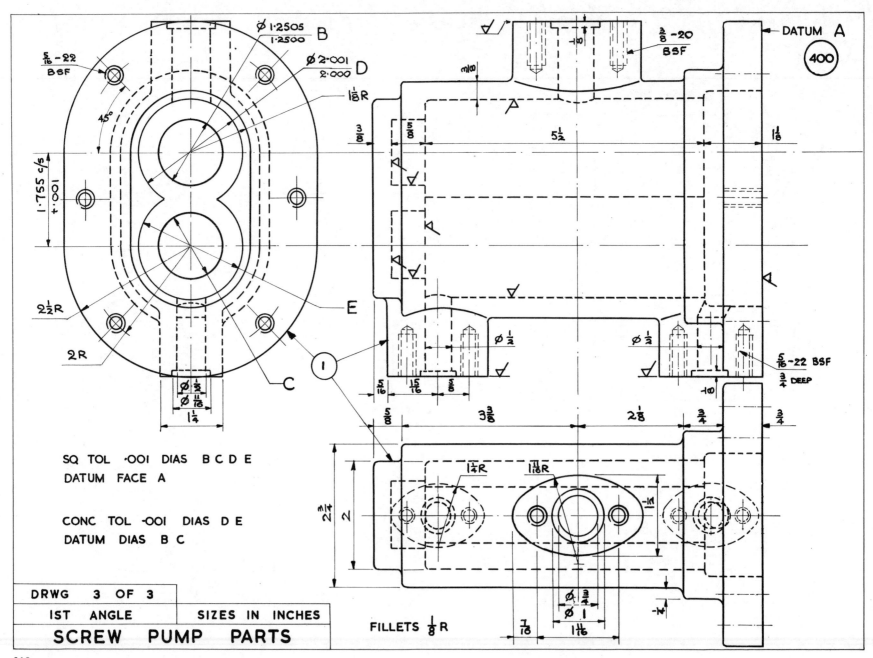

SQ TOL ·OOI DIAS B C D E
DATUM FACE A

CONC TOL ·OOI DIAS D E
DATUM DIAS B C

DATUM A
400

FILLETS ⅛R

398, 399 and 400. Hydraulic Screw Pump The parts of a screw pump are given. The two geared shafts have intermeshing left- and right-hand square threads which carry the fluid from the lower orifices to the delivery port at the top of the casing. The shafts run in bushes pressed in the casing. The upper drive shaft is sealed by a double unit using a sealring and garter spring, the units are held in place by a gland plate.

Draw:

(a) A sectional elevation of the assembled pump taken on the centre line of the shafts, f.s. first angle.

(b) Project a full plan.

(c) Project a full end elevation showing the gland end.

(d) Project a sectional end elevation at the other side of the front elevation taken on centre line of the top orifice.

Add titles and list of parts. Show ten major dimensions.

The worms may be shown by conventional straight lines and not full helical curves; the gears by simple convention.

Part assembly views are given on another page. see 412, also 358.

401. Cylinder Head Two views are shown of a cylinder head, valve, and spring and cover plate. See 421.

1. Draw:

(a) The given left-hand elevation f.s.

(b) Project a plan, from (a) in Third Angle, sectioned on the centre line of the valve.

(c) Project both end elevations, full views.

Add ten major dimensions, and a list of the parts.

2. Draw:

(a) The given elevation as the plan in Third Angle Projection.

(b) Project an elevation from (a).

(c) Project both end elevations, one of which should be sectional to show the valve, seating spring and cover plate details.

Add ten dimensions. Add titles and list of parts.

(d) Make an Isometric Drawing, natural scale, of the assembly, cut away to show the parts.

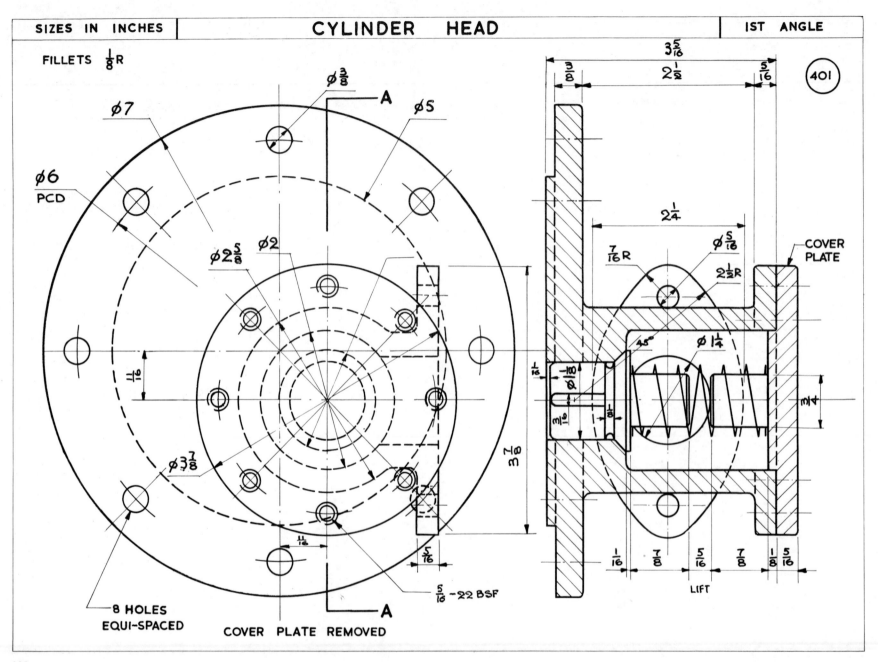

FILLETS ⅛R

∅7

∅6
PCD

∅³⁄₈

A

∅5

∅2⅝   ∅2

∅3⁷⁄₈

11⁄16

5⁄16

5⁄16 -22 BSF

8 HOLES
EQUI-SPACED

COVER PLATE REMOVED

A

3⁵⁄₁₆
2½
5⁄16
3⁄8

(401)

2¼

7⁄16 R

∅5⁄16

2½R

COVER
PLATE

45°   ∅1¼

3⁷⁄₈

3⁄4

1⁄16   7⁄8   5⁄16   7⁄8   1⁄8   5⁄16

LIFT

402. Overhead Swivel Pulley A sectional elevation is given, and a part full end elevation.

Draw: f.s., using First Angle Projection, detail working drawings of the ten separate parts, unassembled.

The vee pulley is free to turn on its shaft, which is held in the fork by two collars and taper pins. The fork is screwed to take a circular retaining collar which is tightened by a peg spanner fitting in the two holes shown. The downward pressure is taken by a radial needle roller caged unit revolving between two hardened steel plates. The whole assembly is fixed by six bolts 8 mm diameter spaced equally on the pitch circle shown.

Show a detailed list of the parts in a suitable position on the drawing.

Show a method of locking the retaining collar.

Draw the assembly as an Isometric Drawing, natural scale, sectioned on the vertical centre line, front quarter removed. See 422.

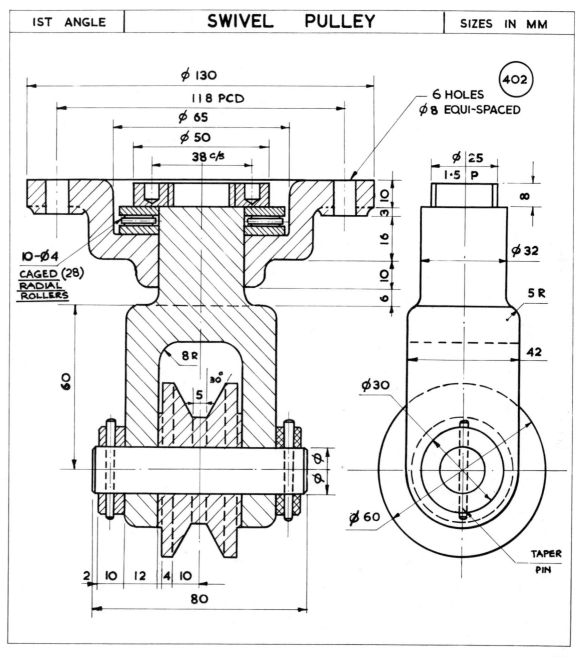

1ST ANGLE	SWIVEL PULLEY	SIZES IN MM

402

6 HOLES
⌀ 8 EQUI-SPACED

⌀ 130
118 PCD
⌀ 65
⌀ 50
38 c/s

10-⌀4
CAGED (28) RADIAL ROLLERS

8 R
30°
5

60

2 10 12 4 10
80

⌀ 25
1·5 P
8
⌀ 32
5 R
42

⌀ 30
⌀ 60

TAPER PIN

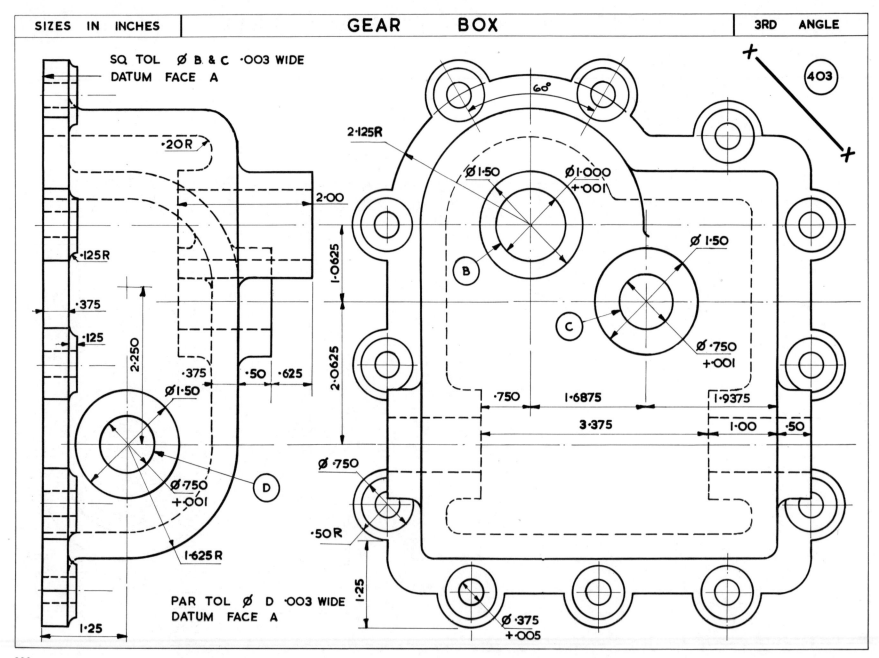

SQ TOL Ø B. & C ·003 WIDE
DATUM FACE A

·20R

·125R

·375

·125

2·250

·375 ·50 ·625

Ø1·50

2·00

1·0625

2·0625

Ø·750 +·001

B

D

1·625R

PAR TOL Ø D ·003 WIDE
DATUM FACE A

2·125R

Ø1·50

Ø1·000 +·001

60°

403

Ø1·50

C

Ø·750 +·001

·750 1·6875 1·9375

3·375

1·00 ·50

Ø·750

·50R

1·25

Ø·375 +·005

403. Gear Box Third Angle Projections of a Gear Box Cover are shown.

(*a*) Take the right-hand view given as the plan for new projections, first angle, f.s., and project a front elevation and an end elevation to the left. The plan and end elevation should be full, but section the front elevation on the centre line of the 1″ diam. shaft.

(*b*) Project a new sectional elevation normal to the line XX. Take the section passing through the centre of the 1″ diam. shaft.
 Dimension the full diagrams (2), add titles.

(*c*) Draw an Isometric View of the box, from the most convenient point.

404, 405, 406 and 407. Swashplate Pump The parts of a swashplate pump are shown in Third Angle Projection. Four plungers are actuated by a face cam, the rams are kept in contact with the face by springs. The shaft runs on a radial loaded ball bearing and a caged needle roller bearing. Four mushroom-headed inlet valves are arranged round a central inlet opening. Four outlet valves are arranged round the periphery of the valve body casting. All the eight valves are lightly spring loaded to help in re-seating.

 Draw:

(*a*) F.S. a sectional front elevation taken on the centre line.

(*b*) Project in First Angle, a full end elevation to the left of the elevation.

(*c*) Project a full plan.

(*d*) Project a sectional end elevation to the right of the front elevation taken on the centre line of the cored holes of the inlet valves.

Calculate the output of the pump when running at 2,000 r.p.m.

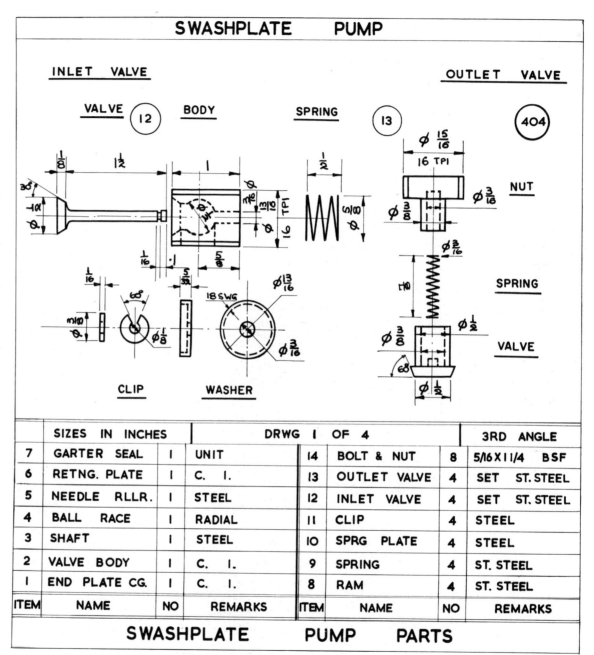

SIZES IN INCHES			DRWG 1 OF 4			3RD ANGLE
7	GARTER SEAL	I	UNIT	14	BOLT & NUT	8	5/16 X 1 1/4 BSF
6	RETNG. PLATE	I	C. I.	13	OUTLET VALVE	4	SET ST. STEEL
5	NEEDLE RLLR.	I	STEEL	12	INLET VALVE	4	SET ST. STEEL
4	BALL RACE	I	RADIAL	11	CLIP	4	STEEL
3	SHAFT	I	STEEL	10	SPRG PLATE	4	STEEL
2	VALVE BODY	I	C. I.	9	SPRING	4	ST. STEEL
1	END PLATE CG.	I	C. I.	8	RAM	4	ST. STEEL
ITEM	NAME	NO	REMARKS	ITEM	NAME	NO	REMARKS

SWASHPLATE PUMP PARTS

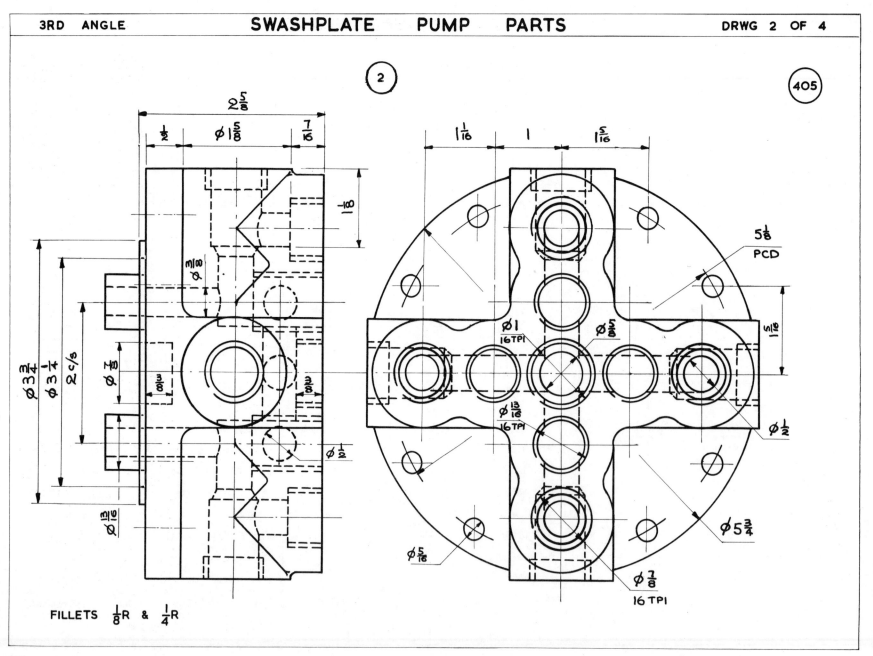

FILLETS $\frac{1}{8}$R & $\frac{1}{4}$R

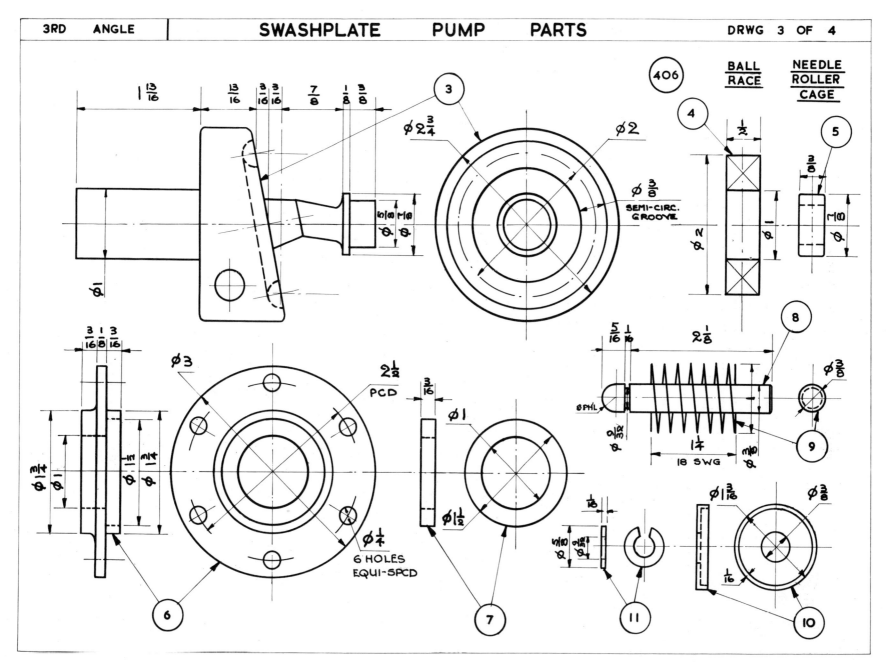

406

BALL RACE

NEEDLE ROLLER CAGE

ø2¾

ø2

ø⅜ SEMI-CIRC. GROOVE

ø3

2½ PCD

6 HOLES EQUI-SPCD

ø¼

18 SWG

SPHL

225

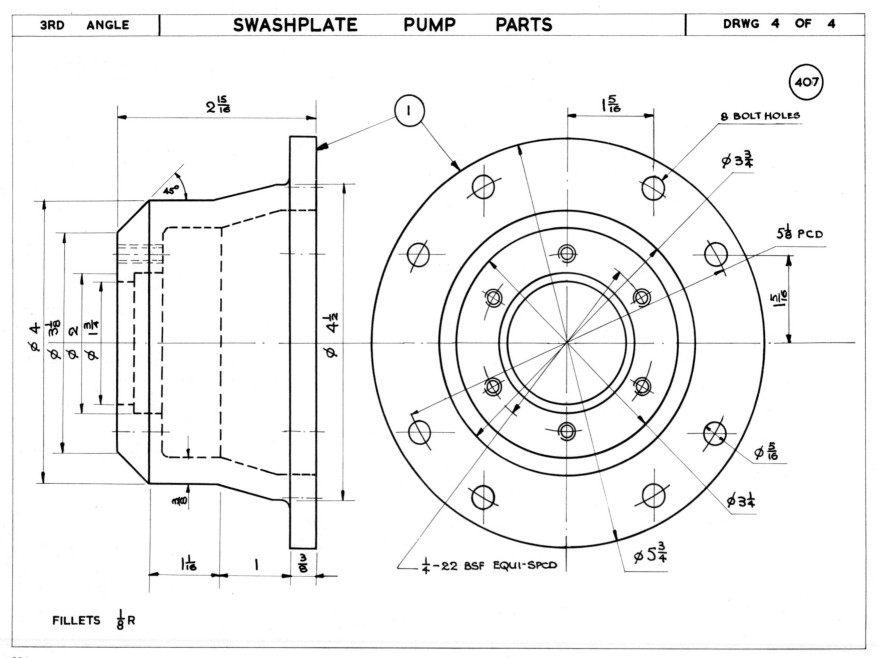

407

$2\frac{15}{16}$

45°

$\phi 4$

$\phi 3\frac{3}{8}$

$\phi 2$

$\phi 1\frac{3}{4}$

$\phi 4\frac{1}{2}$

$\frac{7}{10}$

$1\frac{1}{16}$

1

$\frac{3}{8}$

1

$1\frac{5}{16}$

8 BOLT HOLES

$\phi 3\frac{3}{4}$

$5\frac{1}{8}$ PCD

$1\frac{5}{16}$

$\phi \frac{5}{16}$

$\phi 3\frac{1}{4}$

$\phi 5\frac{3}{4}$

$\frac{1}{4}$-22 BSF EQUI-SPCD

FILLETS $\frac{1}{8}$ R

226

ASSEMBLY SOLUTIONS

Ram Pump

Rotary Vane Pump

Face Valve Pump

Camshaft Pump

Hydraulic Screw Pump

Worm Reduction Gear

Swashplate Pump

Ratchet Feed Gear

ASSEMBLY SOLUTIONS

Orthographic projections in either First or Third Angle are shown of some of the problems set in the previous pages.

408. Ram Pump Assembly   Third Angle. A full plan is shown. A sectional front elevation on the centre line of the ram is projected below the plan.

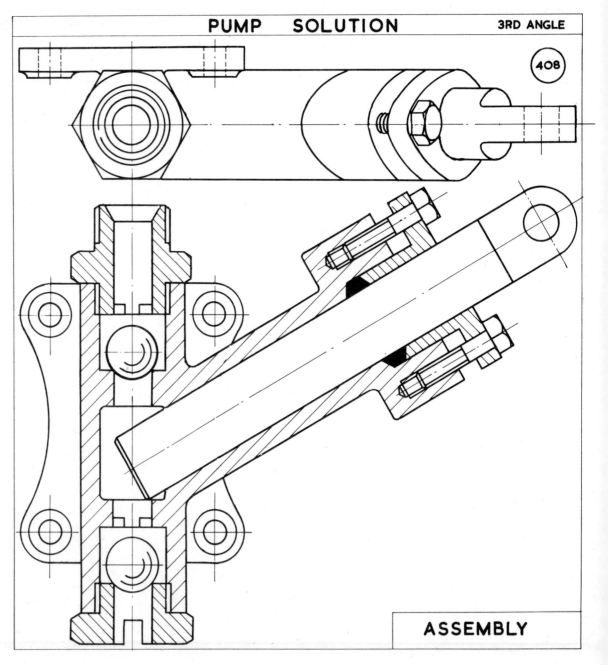

ASSEMBLY

228

409. Vane Pump Assembly Third Angle. A sectional plan is shown taken on the centre line, the rotor is shown in full. The union is shown in full section for ease of drawing.

The front elevation is shown with the front cover removed to display the rotor.

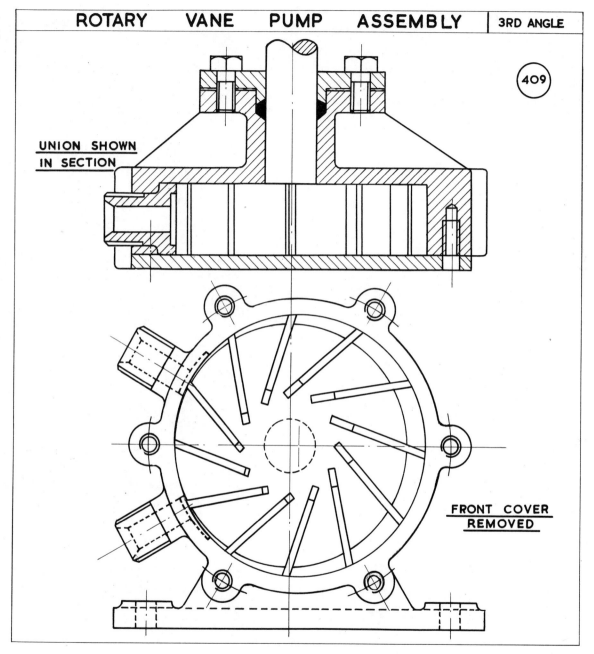

ROTARY VANE PUMP ASSEMBLY | 3RD ANGLE

409

UNION SHOWN IN SECTION

FRONT COVER REMOVED

229

410. Face Valve Pump Cylinder Head Assembly Third Angle. The plan shows the cylinder head removed to show the face valve. The sectional elevation shows the valve in section. The release valve is shown in section also. A scrap plan shows the true shape of the pipe connection face of the cylinder head.

411. Camshaft Pump Assembly First Angle. A sectional front elevation along the axial centre line of the shaft is shown. The camshaft and plungers are shown in the round as are the ball valves.

The sectional end elevation is taken on the centre line of one of the plungers. The cams and plunger are shown in the round.

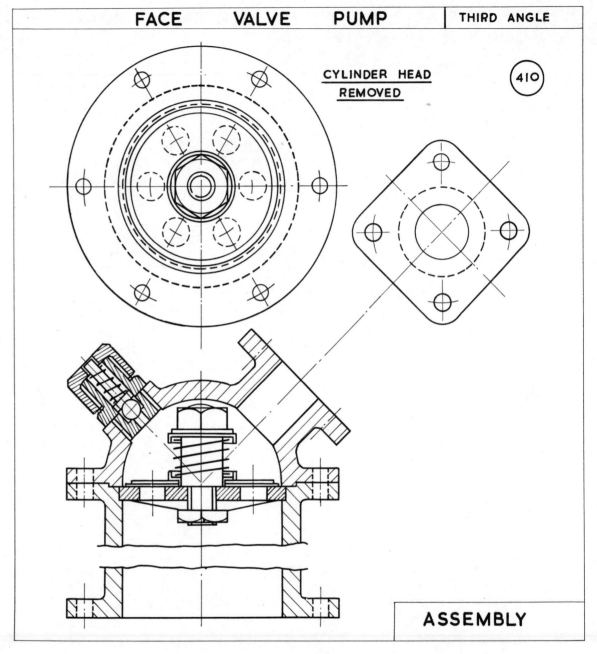

FACE VALVE PUMP | THIRD ANGLE

CYLINDER HEAD
REMOVED

410

ASSEMBLY

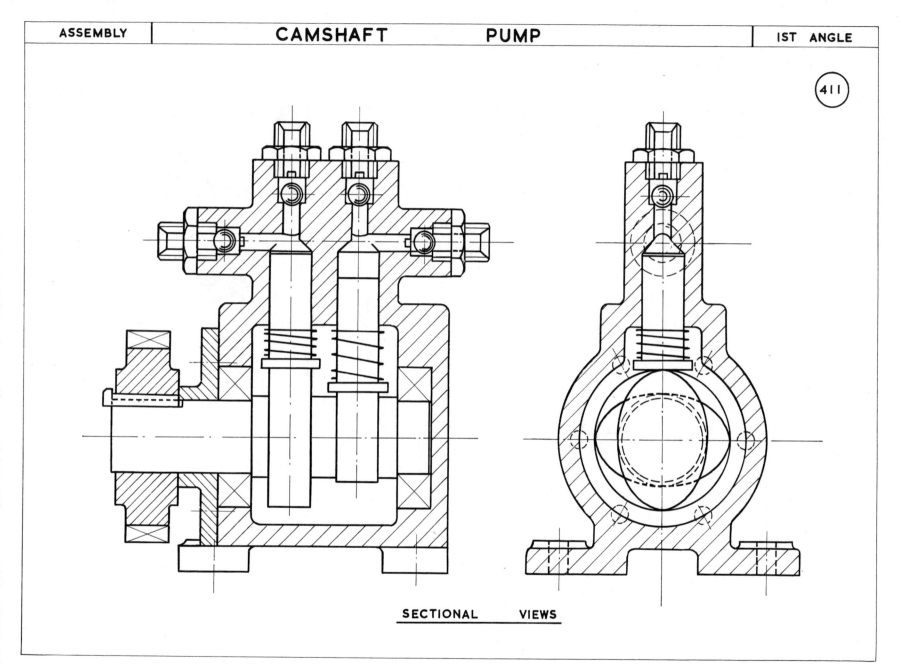

411

SECTIONAL   VIEWS

231

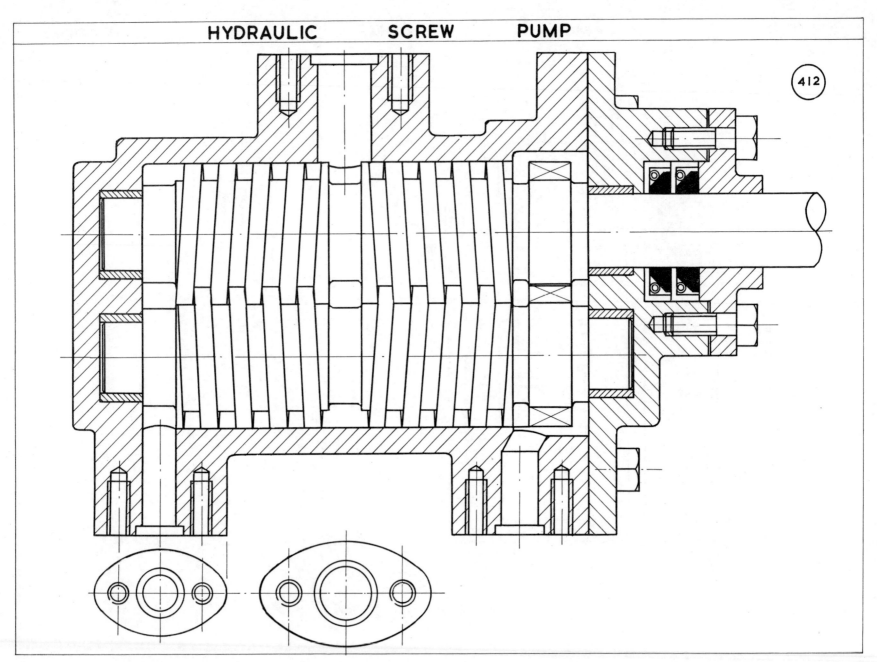

412

ASSEMBLY SOLUTIONS

Orthographic Projections in either First or Third Angle are shown of some of the problems set in the previous pages.

412. Hydraulic Screw Pump Third Angle. A sectional front elevation taken on the centre line of the shafts is shown. The shafts are shown in the round, with the right- and left-hand meshing screwthreads in simple conventional form as are the gears. Scrap details show the shape of the pipe flange coupling and recess for the soft copper seal ring. An end elevation is shown, half as full view, and half sectionalised on the vertical centre line of the outlet.

413. Worm Reduction Gear Assembly First Angle. A sectional front elevation is shown taken on the centre line. The worm is shown simple conventionally in full; three meshing involute teeth of the wormwheel are shown.

The end elevation is sectionalised on the centre line of the wheel and shaft.

414. Swashplate Pump Assembly Third Angle. A sectional front elevation is shown; the swashplate shaft, two rams and two valves are shown in full round form.

A full end elevation is shown.

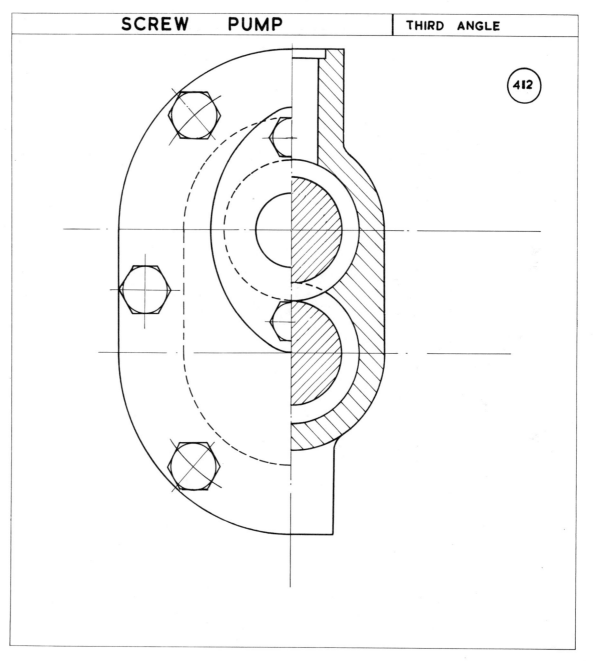

SCREW PUMP | THIRD ANGLE

412

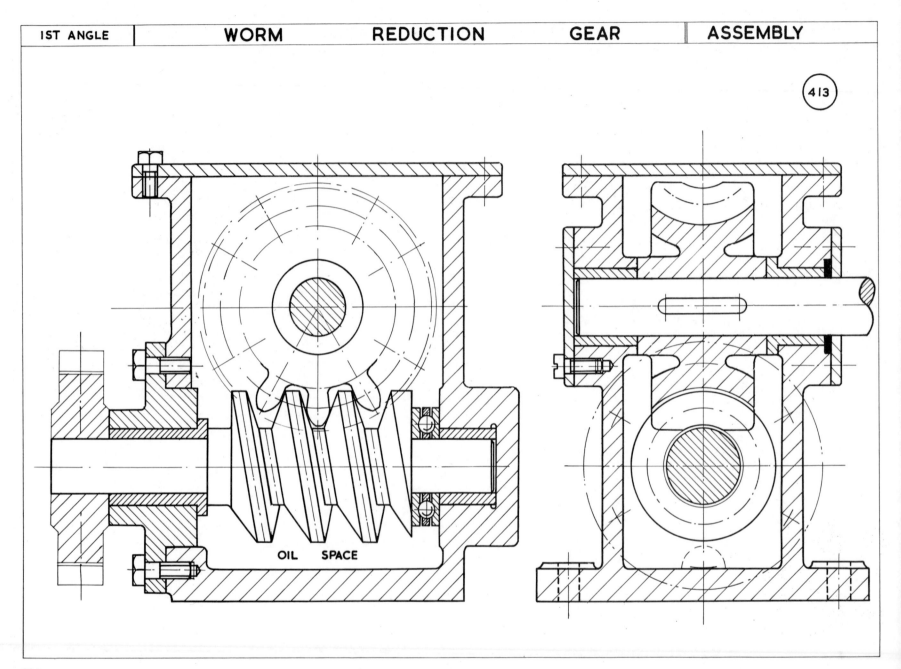

OIL   SPACE

413

234

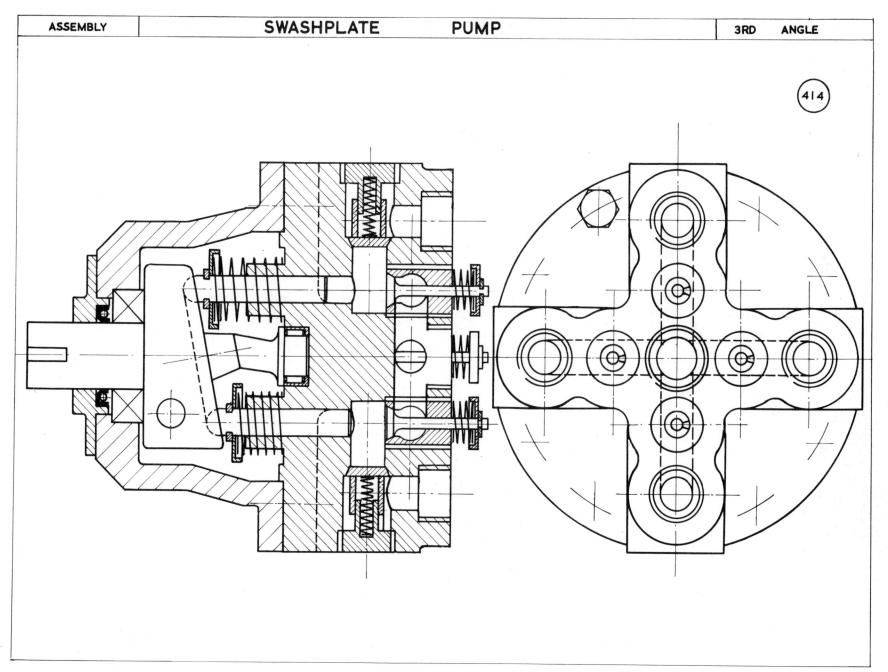

414

414A. Ratchet Feed Gear The system consists of a lever, a spring loaded pawl, and a ratchet wheel. The wheel is fitted to the square section on the shaft, the lever is free to swing on the shaft and its oscillatory motion actuates the pawl which turns the wheel and shaft through a set number of degrees with every stroke. The shaft has spur teeth milled in it, to mesh with a spur gear wheel giving a reduction, and may be used to lift the table of a milling machine in an automatic feed.

Draw:

(a) A sectional front elevation of the assembly, supplying any details which are missing, keeping the proportions similar to the given diagrams.
(b) Project an end elevation, First Angle.
(c) Project a plan from the front elevation. Dimension the drawings, add titles.
(d) Draw detail drawings f.s. Third Angle Projection of the separate parts of the assembly—shaft (teeth need not be shown), lever, ratchet-wheel, washer, nut, pawl, screw and washer, stud and spring, spur gear.

See 423.

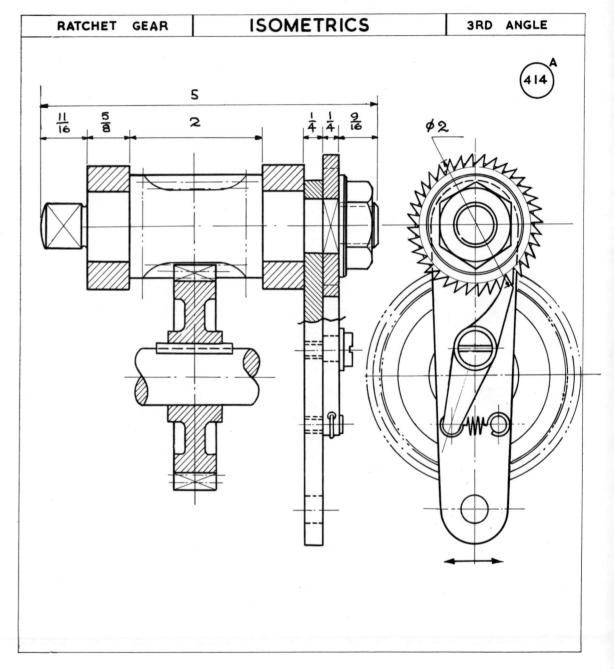

RATCHET GEAR | ISOMETRICS | 3RD ANGLE

236

ISOMETRIC DRAWINGS OF ASSEMBLIES

Roller Bearing

Cone Clutch

Flexible Coupling

Camshaft Pump

Cylinder Head

Swivel Pulley

Ratchet Feed Gear

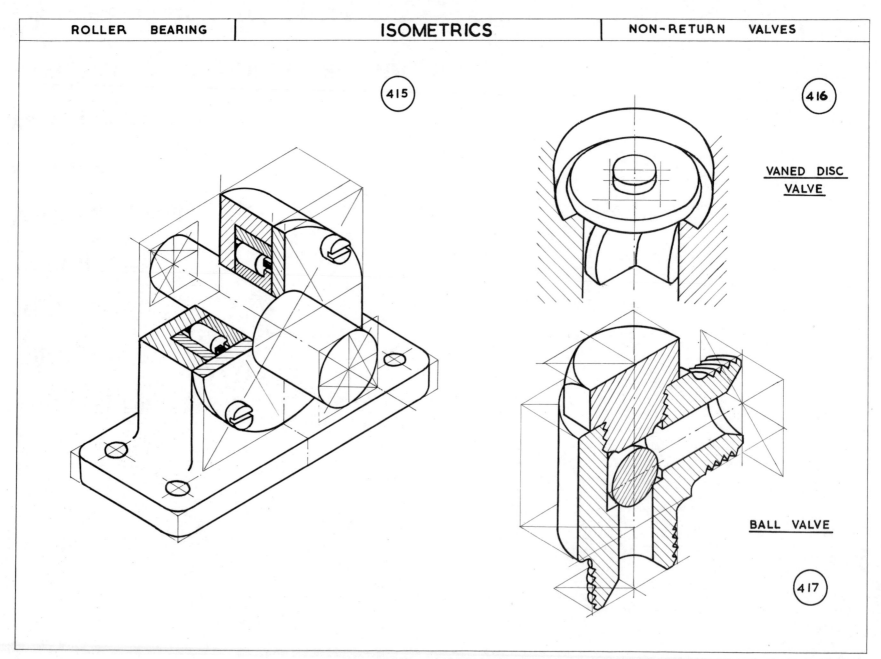

415

416

VANED DISC
VALVE

BALL VALVE

417

238

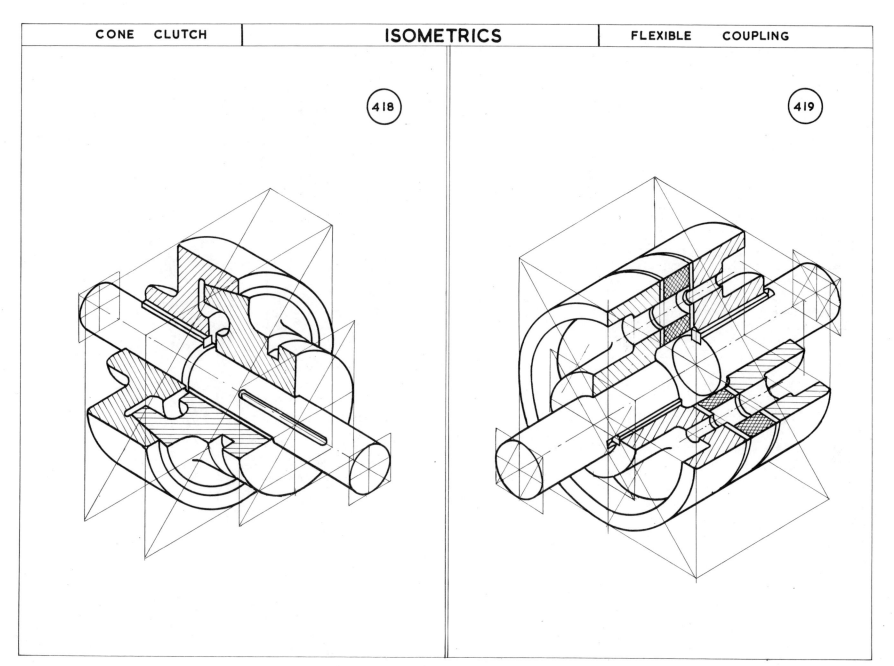

ISOMETRIC DRAWINGS

Isometric Drawings of several examples given on earlier pages, sectionalised to show hidden details.

415. Roller Bearing The diagram shows an Isometric View of the housing, rollers and races, cover plate and shaft. Begin the drawing by drawing the shaft, then the housing and plate as simple 'boxes'. Draw the ellipses using the compass four-arc method. Sectionalise the housing. Draw in the rollers.

416. Disc Valve with Vanes. The vanes are shown with a helical twist which causes the valve to rotate to a fresh position on re-seating.

417. Ball Check Valve A sectionalised valve box, cap and ball valve are shown.

418. Cone Clutch An orthographic section is shown on an earlier page. Draw the shaft first, then the 'boxes' enclosing the two halves of the clutch. Draw the sectioning. Draw the ellipses by the four-arc method.

419. Flexible Coupling Draw the shaft first, followed by the 'boxes' enclosing the coupling discs. Draw the sections. Draw the ellipses by the compass four-arc method.

420. Camshaft Pump Begin by drawing the camshaft. The cam curves are obtained by drawing the enclosing 'box' and ordinates. Draw the section surfaces. Draw 'boxes' for the housing and plate. Draw the ellipses for the housing shapes.

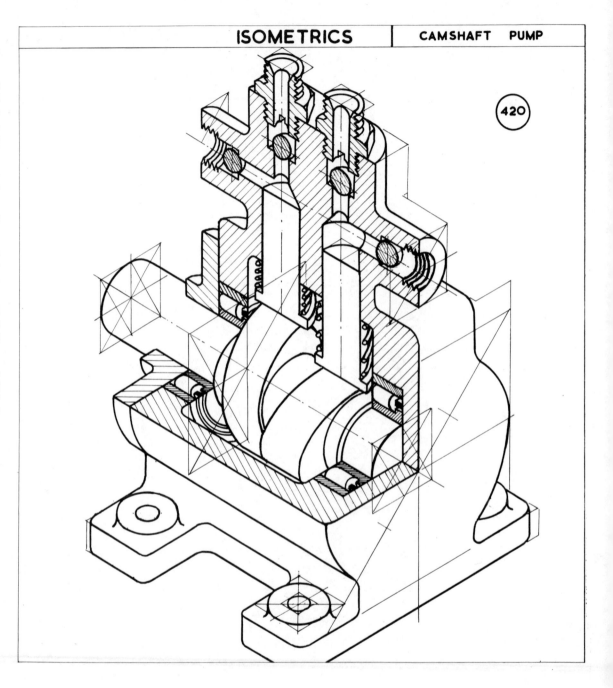

420

240

ISOMETRIC DRAWINGS

421. Cylinder Head and Spring Loaded Valve. A sectionalised Isometric Drawing of the cylinder head showing the vaned valve and spring.

422. Swivel Pulley A sectionalised Isometric Drawing of the swivel pulley showing the parts assembled. Draw the top plate first. Draw the section surfaces. Draw ellipses for the shaft and retaining collars.

423. Ratchet Gear A sectionalised Isometric Drawing of the ratchet feed gear. Begin by drawing the shaft. Draw the section surfaces. Draw 'boxes' for bearings, lever, ratchet wheel and nut. Obtain the final shapes by ordinates and ellipses. Draw the pawl, screw, stud and spring. Enclosing 'boxes' will assist in arriving at the final shapes.

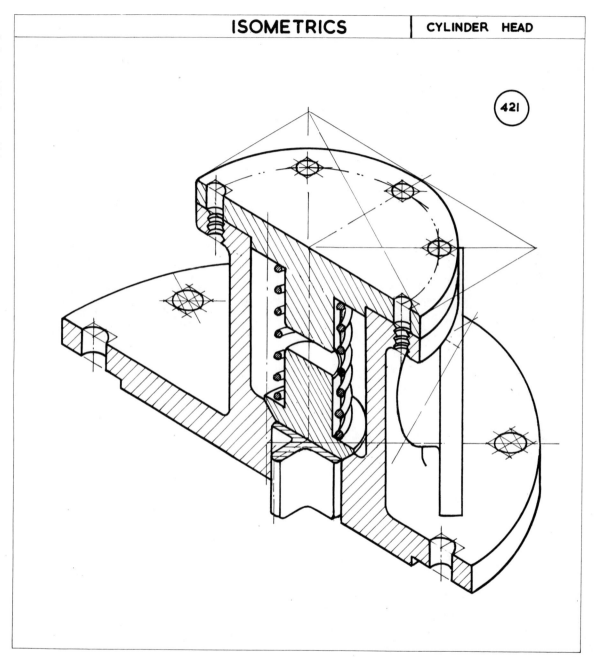

421

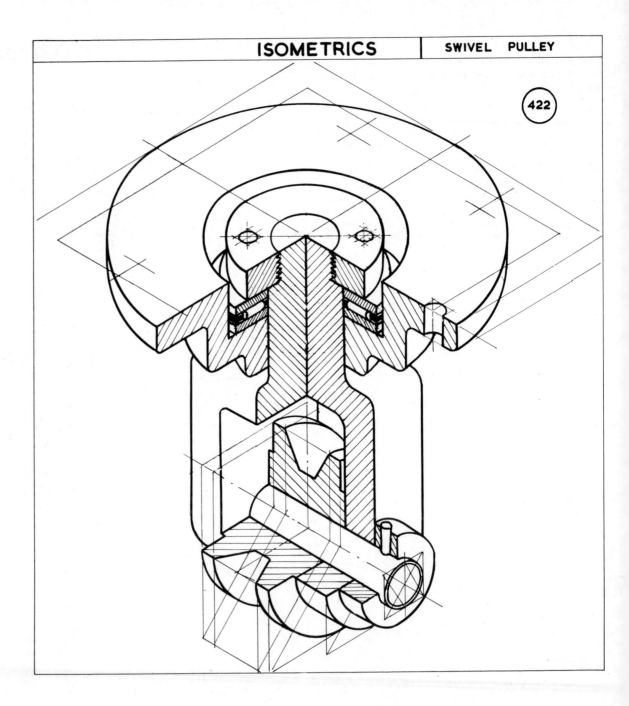

| ISOMETRICS

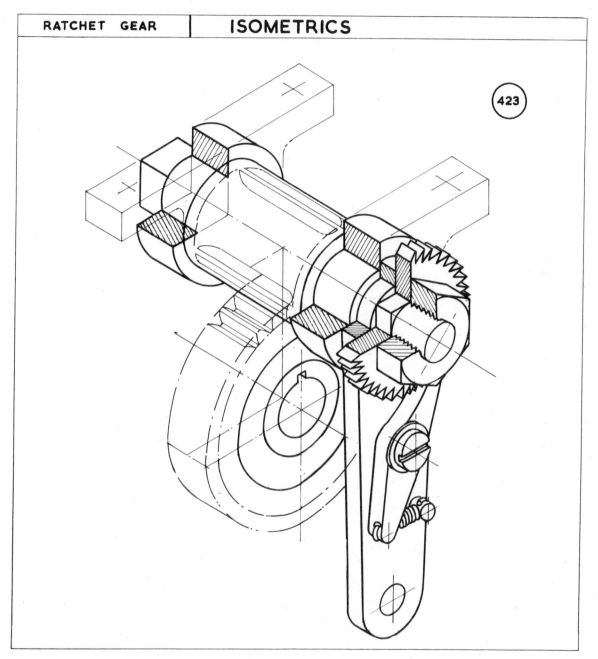

423

ELECTRICAL CIRCUITS

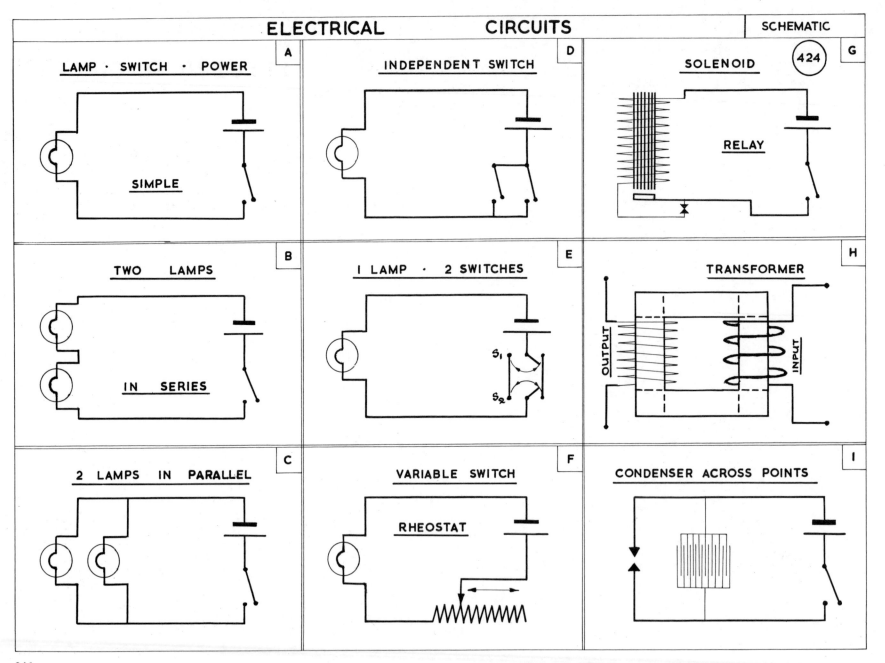

A — LAMP · SWITCH · POWER — SIMPLE

B — TWO   LAMPS — IN   SERIES

C — 2   LAMPS   IN   PARALLEL

D — INDEPENDENT SWITCH

E — 1 LAMP · 2 SWITCHES — S_1 S_2

F — VARIABLE SWITCH — RHEOSTAT

G — SOLENOID — 424 — RELAY

H — TRANSFORMER — OUTPUT — INPUT

I — CONDENSER ACROSS POINTS

424.

A. Simple circuit of lamp, battery and control switch.

B. Two lamps in series. The current is reduced by the equal lamps which glow at half power.

C. Two lamps in parallel, the two lamps glow at full illumination, double current being taken from the battery.

D. One lamp controlled by either of two switches.

E. One lamp controlled by either switch.

F. Variable switch (rheostat). One lamp controlled from zero to full illumination by a slider contact on a resistance coil.

G. Soft iron core magnetised by the field in the coil when the switch is closed. The armature is attracted and closes a switch to operate a further circuit.

H. Two coils wound round a common armature made up of silicon iron laminations, insulated. The input and output are in a ratio of the number of turns in each. No connection between the two coils; the electromotive force (e.m.f.) is induced from the input primary winding into the output secondary winding.

I. Tungsten or silver points which make and break are prevented from pitting by sparking by placing a foil and insulation condenser across the points.

425.

J. Heater element consisting of a resistance wire—nickel-chrome—wrapped round a porcelain tube.

K. Carefully balanced armature working between the poles of a permanent magnet deflects when a current is passed.

L. Coil ignition circuit commonly used on a four-cylinder automobile engine. 12-volt primary circuit induces a high-tension current in the secondary winding to cause an ignition spark at the plug. A condenser is placed across the points to accumulate the electric charge and intensify the spark. The four lift cam opens the contact points.

M. Wound field and armature, in series. The current passes through the armature, thence through the field.

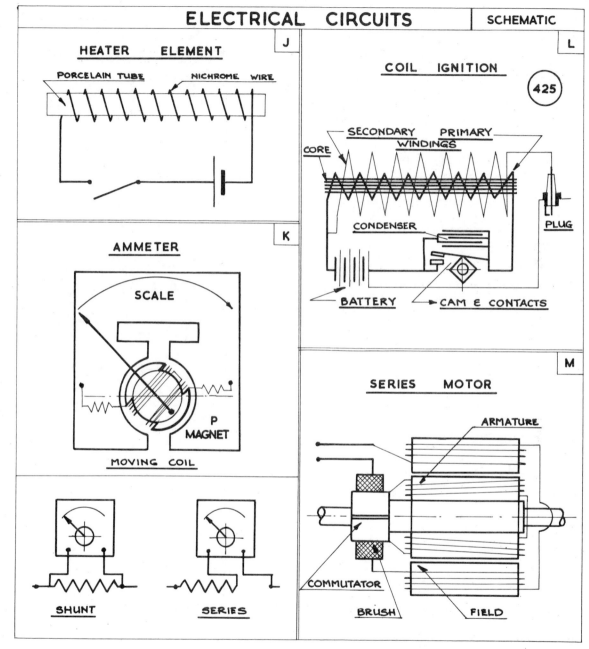

ELECTRICAL CIRCUITS — SCHEMATIC

426.

A. Diagram for a shunt wound motor. The field and armature are connected in parallel.

B. Diagram showing a motor wound in series. The current passes through the armature and then through the field.

C. Starter and field windings are connected as shown.

D. Electro-magnetic field attracts a simple polar armature. The rotation of the rotor is dependent on the number of poles and the cycle frequency of the alternating current. The rotor must receive a speed impulse on starting.

E. Direct current motors require a resistance starter to prevent current becoming excessive when starting. A solenoid holds the switch in the running position once the normal motor speed has been attained.

F. The diagram shows a circuit used in the electrolytic deposition of metals. The system is also used in the shaping of metal by wasting electrolytically. Metal is wasted away from the anode and deposited on the cathode as the current flows from positive to negative.

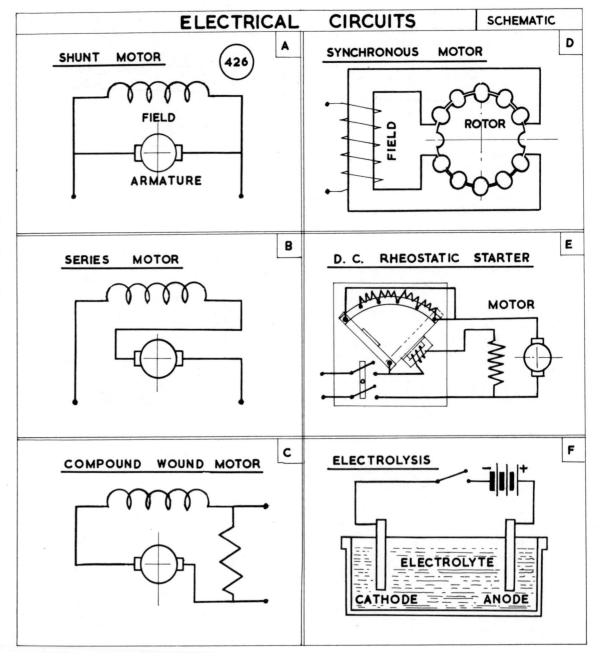

427.

A. **Capacitor Start**, split phase electric motor. In many smaller type domestic single phase motors, the stator is laminated and has a field winding in series for normal running. A second field winding with a capacitor or large condenser aids in starting. After normal running speed is attained, a centrifugal switch cuts out the starter coil.

B. **Contactor Motor Switch** A switch operated by a solenoid or electro-magnet. When the ON button is pressed, current passes through the solenoid coil and the armature is attracted closing the supply line contacts to pass current to the motor. If the system overheats due to overloading, the bi-metal strips at $D_{1,2,3}$, are heated by resistance wires, and expand to operate an overload trip switch to break the circuit, saving damage to the equipment. The motor is stopped in the normal way by pressing the OFF button which breaks the solenoid circuit releasing the armature and line contacts.

Most contactors are of the airbreak type with silver contacts; oil immersed contacts to avoid arcing, are used for large motors.

C. **Telephone Circuit** The telephone consists of a microphone and a receiver connected by wiring and energised by battery or low power mains.

The microphone or transmitter consists of a steel diaphragm backed by a capsule of carbon granules which pass varying amounts of current dependent on the pressure on the diaphragm by the sound waves. The receiver consists of a stalloy diaphragm which is made to vibrate and cause sound waves in tune with the microphone. The varying current passes through the coils of the electro-magnet in the receiver causing a varying magnetic field and thus vibrating the diaphragm.

A transformer in the circuit boosts operation.

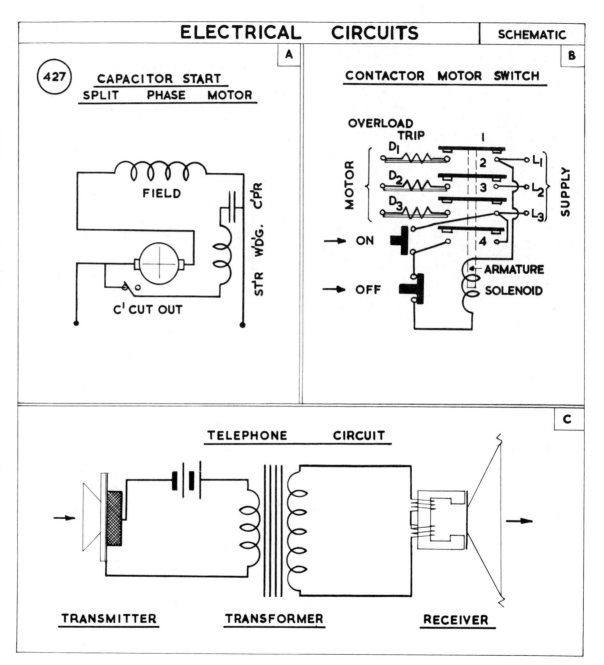

428. Electrical Layouts

A. **Thermostats** A. Bi-metal Strip. The bi-metal strip is made from two strips of metal having widely different coefficients of expansion. As the strip heats up, bending takes place opening the contacts and breaking the circuit.

B. Invar Rod. A brass tube holds an Invar rod. As the tube heats up and expands at a greater rate than the rod the contacts open.

C. A fluid-filled sealed capsule expands on heating to open the contacts.

B. **Magnetic Separation** Mixed metallic scrap passes down the chute, the ferrous scrap is drawn further round the revolving drum to drop in the second hopper. A continuous magnetic field holds the iron scrap to the drum.

C. **Lifting Magnets** A motor driven D.C. generator supplies current to a large electro-magnet held by a crane and is used to carry ferrous scrap. The current passes through a voltage regulator V and a contactor switch C. The current may be reversed to speed up the drop of the scrap after transport. As an example a 30″ diam. magnet working on 220 volts D.C. 13·9 amps will lift 13,000 lb.

D. In metallic arc welding, a low voltage—25 to 50 volts—and high amperage D.C. current is passed through a welding electrode and the object to be welded. The arc temperature melts the electrode and enables welded or fused joints to be made.

The diagram shows a circuit for a cross field series generator with a variable resistance.

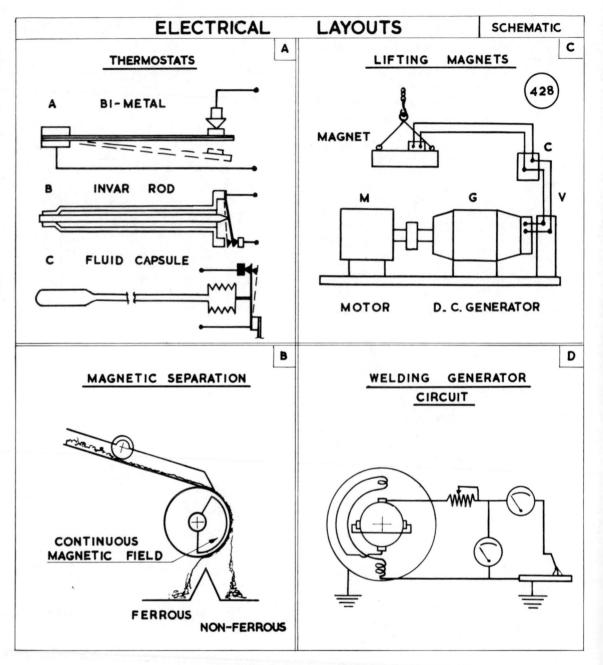

THERMOSTATS

A BI-METAL

B INVAR ROD

C FLUID CAPSULE

LIFTING MAGNETS

428

MAGNET

M   G   V

C

MOTOR   D. C. GENERATOR

MAGNETIC SEPARATION

CONTINUOUS MAGNETIC FIELD

FERROUS   NON-FERROUS

WELDING GENERATOR CIRCUIT

429. Generation of Electricity Electricity is generated by three phase, 50 cycle alternators driven by steam turbines. In the latest nuclear power stations, the high pressure (up to 3,500 lb/in^2) steam obtained from the cooling of the uranium reactors, is used to produce A.C. current at 275,000 volts and this is fed into the Super Grid system. Transformer sub-stations convert the current to 132,000 volts which is then fed into the National Grid system being conveyed by the familiar overhead wire and pylon method. Electricity is also generated by steam turbines, the steam for which is obtained from coal fired boilers. These run on much lower pressures 600 to 900 lb/in^2 and the current is 11,000 volts. This is stepped up by transformer sub-stations to 132,000 volts before being accepted by the National Grid.

Electricity is also generated by water power. The water is piped from highland reservoirs to water turbines (Pelham wheels) which drive A.C. alternators supplying current at 11,000 volts requiring transforming before feeding into the grid.

The diagram shows details of the conversion of the power supply from high pressure voltage carried by the grid, to voltages suitable for traction (D.C.); three phase 415 volts for industrial uses; domestic and commercial, 240 volts single phase.

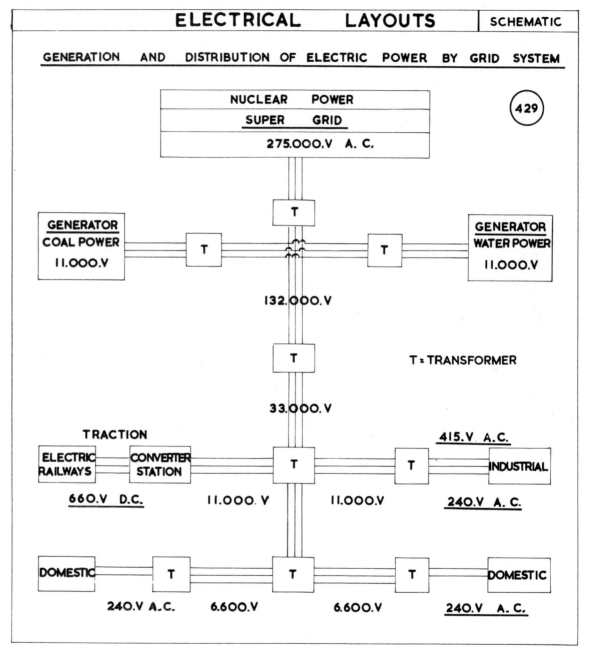

ELECTRICAL LAYOUTS | SCHEMATIC

GENERATION AND DISTRIBUTION OF ELECTRIC POWER BY GRID SYSTEM

NUCLEAR POWER
SUPER GRID
275.000.V A.C.

(429)

T

GENERATOR
COAL POWER
11.000.V

T

T

GENERATOR
WATER POWER
11.000.V

132.000.V

T

33.000.V

T = TRANSFORMER

TRACTION

415.V A.C.

ELECTRIC RAILWAYS

CONVERTER STATION

T

T

INDUSTRIAL

660.V D.C.   11.000.V   11.000.V   240.V A.C.

DOMESTIC

T

T

T

DOMESTIC

240.V A.C.   6.600.V   6.600.V   240.V A.C.

430. Domestic and Commercial Wiring   The street main is usually 415 volts A.C., 50 cycles, three phase, carried on three conductors and a fourth neutral wire.

Two conductors—one phase wire and the neutral—are brought to the house meter through a Supply Board sealed fuse box, giving 240 volts A.C., 50 cycles, single phase current.

From the meter the lights circuits have a common main switch and then individual fuses and circuits controlled by single pole switches on the 'live' wire.

Ring Main   Power points are arranged on a ring circuit which consists of the phase 'live' wire and the neutral wire on a main ring switch and fuse, as shown in the diagram. Power points are then tapped across the two conductors at the required places without the need for long individual wiring back to the fuse box.

Fused plugs are used with each piece of apparatus to avoid overloading.

A local earth wire forms the third wire in the ring circuit, and short circuits the current to earth should accidental breakdown of the insulation take place, providing an additional safeguard to the operator.

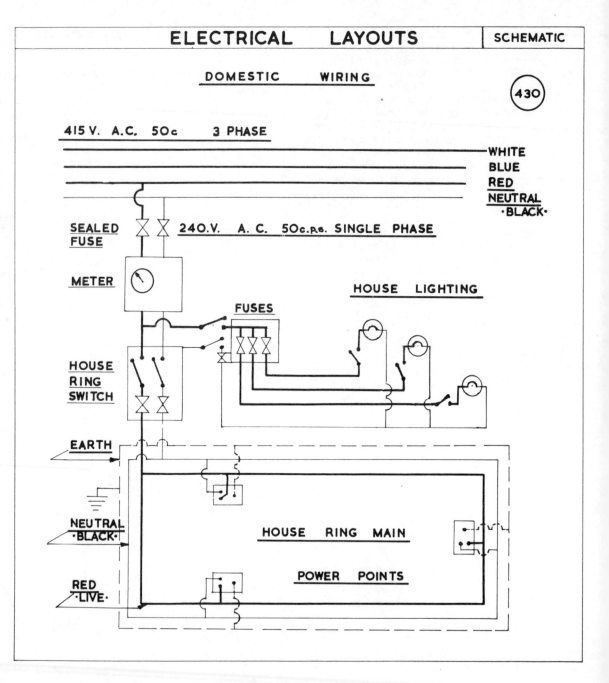

ELECTRICAL   LAYOUTS   SCHEMATIC

DOMESTIC   WIRING

430

415 V. A.C. 50c   3 PHASE

WHITE
BLUE
RED
NEUTRAL
·BLACK·

SEALED FUSE   240.V. A. C. 50c.p.s. SINGLE PHASE

METER   HOUSE LIGHTING

FUSES

HOUSE RING SWITCH

EARTH

NEUTRAL ·BLACK·

HOUSE RING MAIN

POWER POINTS

RED ·LIVE·

FREEHAND SKETCHING

431. Freehand Sketching Freehand sketching without instruments should be practised constantly. The simple technique of outline and shading of a sketch based on an isometric or oblique drawing of simple geometric solids should be attempted first. The object is usually illuminated from the left, and shadow shading effected by hatching lines in two directions on plane surfaces. The brightest effect is obtained by placing the deepest shadow adjacent to the highest light. Analyse the object into its basic geometric solid forms, and sketch boldly to obtain the outline. Pencil shading will give a greater graduation than the pen and ink diagrams shown; HB to 4B pencils can be used.

432. Orthographic pencil sketches are often required, and squared paper allowed. Plans and elevations, first or third angle, sectional to show hidden detail, based on B.S. 308, dimensioned, are shown in the diagrams.

433. More complicated freehand sketches are shown on the following pages, sectionalised orthographic and isometric diagrams are given, car components, lathe, drilling machine, power saw, bench shears, and hand tools of the workshop provide examples for further subjects. Electrical and laboratory gear should also be sketched. List of suggestions is appended.

434. Given an isometric drawing or a photograph of a casting or engineering component, with one dimension supplied, make orthographic drawings full size or to scale on squared paper.

Two castings are shown and one dimension given for each. Draw two elevations and a plan of each of the castings, the first in Third Angle, and the second in First Angle. Dimension the drawings.

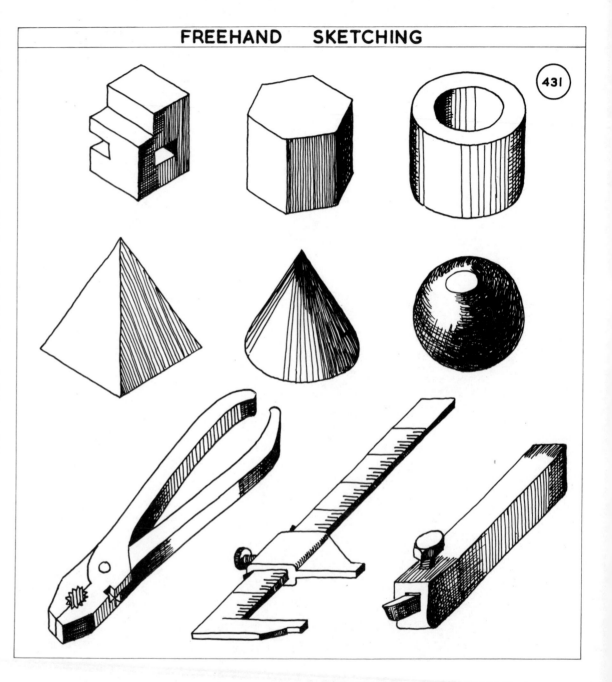

431

FREEHAND   SKETCHING

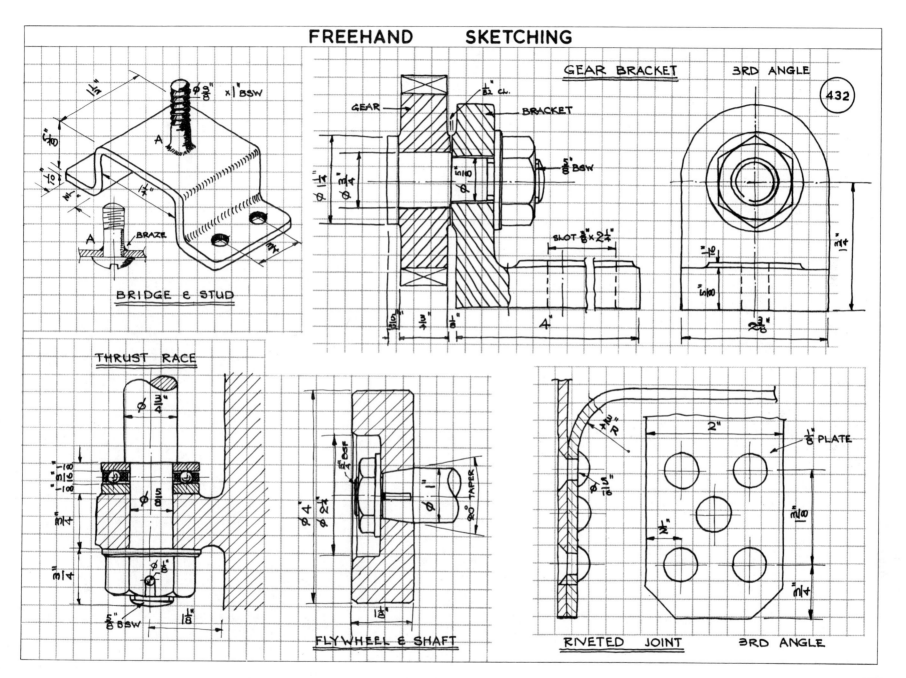

GEAR BRACKET   3RD ANGLE

(432)

BRIDGE & STUD

THRUST RACE

FLYWHEEL & SHAFT

RIVETED JOINT   3RD ANGLE

FREEHAND SKETCHING

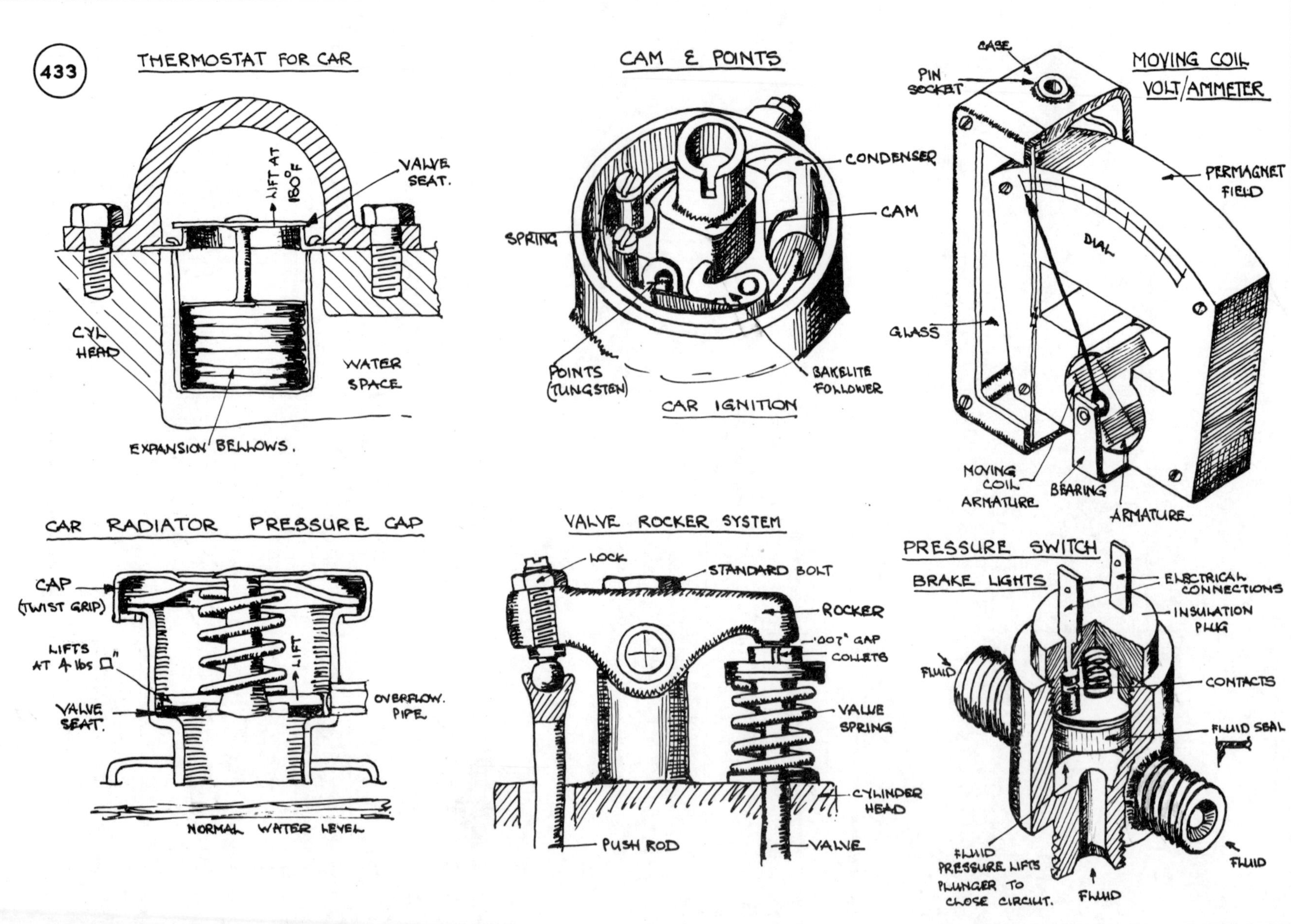

(433)

THERMOSTAT FOR CAR

LIFT AT 160°F
VALVE SEAT.
CYL HEAD
WATER SPACE
EXPANSION BELLOWS.

CAM & POINTS

CONDENSER
CAM
SPRING
POINTS (TUNGSTEN)
BAKELITE FOLLOWER
CAR IGNITION

MOVING COIL VOLT/AMMETER

CASE
PIN SOCKET
PERMAGNET FIELD
DIAL
GLASS
MOVING COIL ARMATURE
BEARING
ARMATURE

CAR RADIATOR PRESSURE CAP

CAP (TWIST GRIP)
LIFTS AT 4 lbs □"
LIFT
VALVE SEAT.
OVERFLOW PIPE.
NORMAL WATER LEVEL

VALVE ROCKER SYSTEM

LOCK
STANDARD BOLT
ROCKER
'007" GAP
COLLETS
VALVE SPRING
CYLINDER HEAD
PUSH ROD
VALVE

PRESSURE SWITCH

BRAKE LIGHTS
ELECTRICAL CONNECTIONS
INSULATION PLUG
FLUID
CONTACTS
FLUID SEAL
FLUID
FLUID PRESSURE LIFTS PLUNGER TO CLOSE CIRCUIT.
FLUID

256

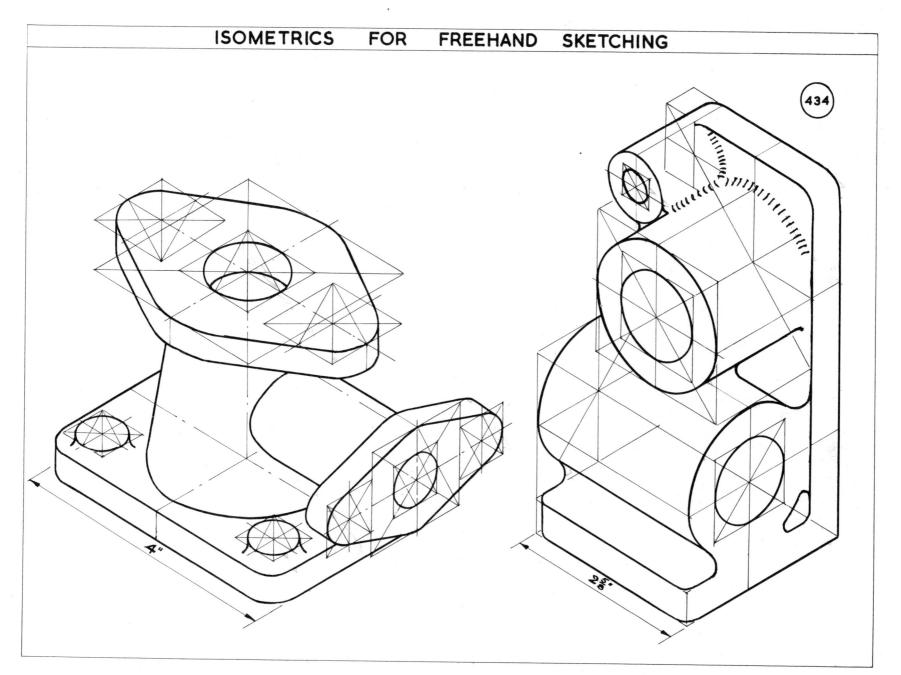

434

4"

2⅜"

SUGGESTIONS FOR SKETCHING

Electrical

Light switch. Circuits 424–430.
15 amp fused plug.
15 amp switch and socket.
Pendant lamp holder.
Batten lamp holder.
Fuse box, fuses.
Mains flow meter.
Venner-type time switch.

Motors

Synchronous clock type.
Battery type impulse motor.
Fractional H.P. series wound commutator type.
Squirrel cage motor.
Centrifugal switch.
Rheostat and switch.
Contactor, solenoid switch.

Car and Motor-cycle parts

Battery section.
Starter motor.
Dynamo.
Ammeter.
Dynamo solenoid cut-out.
Ignition coil and circuit.
Contact breaker and condenser.
Valve gear, push rods, rocker and valves with springs.
Cam shaft.
Oil pump, filter system.
Pistons and crankshaft.
Clutch, gear box schematic.
Universal coupling.
Rear axle, schematic.

Rear spring and axle mounting.
Front wheel bearing.
Brake drums.
Brake master cylinder.
Door catch.
Bonnet catch.
Brake, hand ratchet system.
Wiper motor, rotary to oscillating movement.
Flash indicator mechanism.
Dampers and action.
Brake light switch.
Silencer.
Engineering parts.
Parts of drilling machine.
Parts of lathe.
Lathe processes:
Centring in chuck,
  facing,
    drilling and reaming,
    tapping,
    tailstock
      die threading,
    screwcutting,
    spot-facing,
    milling,
    end-milling,
    keyway cutting,
    taper and steady
      turning.
Component parts 294–357,
  bearings,
  seals,
  couplings,
  flanges,
  brackets and clamps,
riveted joints, 281–293,
bolted joints,
nut locking devices.

258

APPENDIX I   TABLES

Size (in.)	T.P.I.	B.S.W. threads — Effective Diam. (in.)	Minor Diam. (in.)	B.S.F. threads — T.P.I.	Effective Diam. (in.)	Minor Diam. (in.)
$\frac{1}{8}$	40	0·1090	0·0930			
$\frac{1}{4}$	20	0·2180	0·1860	26	0·2254	0·2008
$\frac{5}{16}$	18	0·2769	0·2413	22	0·2834	0·2543
$\frac{3}{8}$	16	0·3350	0·2950	20	0·3430	0·3110
$\frac{1}{2}$	12	0·4466	0·3932	16	0·4600	0·4200
$\frac{5}{8}$	11	0·5668	0·5086	14	0·5793	0·5336
$\frac{3}{4}$	10	0·6860	0·6220	12	0·6966	0·6432
1	8	0·9200	0·8400	10	0·9360	0·8720
$1\frac{1}{2}$	6	1·3933	1·2866	8	1·4200	1·3400
2	$4\frac{1}{2}$	1·8577	1·7154	7	1·9085	1·8170

For further details refer to B.S. 84:1956 (Now superseded by ISO and Unified screw forms.)

Tapping Size, B.S.W., B.S.F. $T = D - 1\cdot1328p$

$$p = \text{pitch} = \frac{1}{\text{t.p.i.}}.$$

Diameter of Rivets:

$$D = 1\cdot2\sqrt{\text{Thickness of plate.}}$$

$1'' = 25\cdot4$ mm

1 mm $= 0\cdot03937''$.

Fraction to Decimal

Fraction (in.)	Decimal (in.)	mm
$\frac{1}{64}$	0·015625	0·39
$\frac{1}{32}$	0·03125	0·79
$\frac{1}{16}$	0·0625	1·59
$\frac{1}{8}$	0·125	3·17
$\frac{3}{16}$	0.1875	4·76
$\frac{1}{4}$	0·250	6·35
$\frac{1}{2}$	0·500	12·70
$\frac{3}{4}$	0·750	19·05
1	1·000	25·40

Unified Thread (size in inches)

Nom. Diam.	Coarse — Designation	T.P.I.	Bolt Root Diam.	Nut Core Diam.	Fine — Designation	T.P.I.	Bolt Root Diam.	Nut Core Diam.
$\frac{1}{4}$	$\frac{1}{4}$-20 UNC	20	0·1887	0·1959	$\frac{1}{4}$-28 UNF	28	0·2062	0·2113
$\frac{5}{16}$	$\frac{5}{16}$-18 UNC	18	0·2443	0·2524	$\frac{5}{16}$-24 UNF	24	0.2614	0·2674
$\frac{3}{8}$	$\frac{3}{8}$-16 UNC	16	0·2983	0·3073	$\frac{3}{8}$-24 UNF	24	0·3239	0·3299
$\frac{1}{2}$	$\frac{1}{2}$-13 UNC	13	0·4056	0·4167	$\frac{1}{2}$-20 UNF	20	0·4387	0·4459
$\frac{5}{8}$	$\frac{5}{8}$-11 UNC	11	0·5135	0·5266	$\frac{5}{8}$-18 UNF	18	0·5568	0·5649
$\frac{3}{4}$	$\frac{3}{4}$-10 UNC	10	0·6273	0·6417	$\frac{3}{4}$-16 UNF	16	0·6733	0·6823
1	1-8 UNC	8	0·8466	0·8647	1-12 UNF	12	0·8978	0·9098
$1\frac{1}{2}$	$1\frac{1}{2}$-6 UNC	6	1·2955	1·3196	$1\frac{1}{2}$-12 UNF	12	1·3978	1·4098
2	2-$4\frac{1}{2}$ UNC	$4\frac{1}{2}$	1·7274	1·7594				

For further details refer to B.S. 1580:1962.

B.A. Screwthreads (sizes in mm)

No.	Diam.	Pitch	Eff. Diam.	Minor Diam.
0	6·0	1·00	5·40	4·80
2	4·7	0·81	4·21	3·73
4	3·6	0·66	3·20	2·81
6	2·8	0·53	2·58	2·16
8	2·2	0·43	1·94	1·68
10	1·7	0·35	1·49	1·28
12	1·3	0·28	1·13	0·96

For further details refer to B.S. 93:1951.

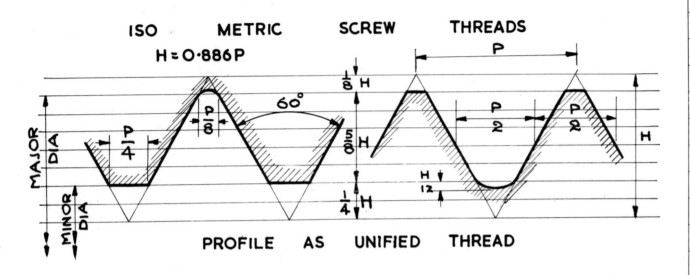

ISO METRIC SCREW THREADS

$H \approx 0.886P$

60°

MAJOR DIA

MINOR DIA

$\frac{P}{4}$ $\frac{P}{8}$ $\frac{1}{8}H$ $\frac{5}{8}H$ $\frac{1}{4}H$

P $\frac{P}{2}$ $\frac{P}{2}$ $\frac{H}{12}$ H

PROFILE AS UNIFIED THREAD

ISO Metric Threads (sizes in mm)		
Major Diam.	**Minor Diam.**	**Pitch**
1·6	1·22	0·35
2	1·57	0·4
2·5	2·01	0·45
3	2·46	0·5
4	3·24	0·7
5	4·13	0·8
6	4·92	1·0
8	6·65	1·25
10	8·38	1·5
12	10·11	1·75
16	13·83	2·0
20	17·29	2·5
24	20·75	3·0
30	26·21	3·5
36	31·67	4·0

For further details refer to B.S. 3643: Part 1: 1963.

Scales	Representative Fraction	
Full size	1/1	F.S.
Half size	1/2	(6″ = 12″).
One-third	1/3	(4″ = 12″).
Quarter size	1/4	(3″ = 12″).
One-sixth	1/6	(2″ = 12″).
One-eighth	1/8	(1½″ = 12″).
One-twelfth	1/12	(1″ = 12″).

Small Scales
$\frac{3}{4}″$ = 1 ft 1/16. F.S.
$\frac{1}{2}″$ = 1 ft 1/24. F.S.
$\frac{3}{8}″$ = 1 ft 1/32. F.S.
$\frac{1}{4}″$ = 1 ft 1/48. F.S.
$\frac{1}{8}″$ = 1 ft 1/96. F.S.

Selection of British Standards

B.S. 10. Flanges for Pipes, Valves and fittings.
- 21. Pipe Threads.
- 46. Part I. Keys and Keyways.
- 57. B.A. Bolts, Nuts and Screws.
- 84. Whitworth Threads.
- 93. B.A. Threads.
- 292. Ball and Roller Bearings.
- 308. Engineering Drawing Practice.
- 351. Friction Surface, Rubber Transmission Belting.
- 436. Helical and Spur Gears.
- 545. Bevel Gears.
- 641. Small Rivet Dimension.
- 721. Worm Gearing.
- 1440. Endless Belt Drives.
- 1580. Unified Threads.
- 1916. I.E.I. Guide to Selection of Fits, Limits and Tolerances.
- 2035. C.I. Flanged Pipe and Fittings.
- 2059. Splines and Serrations.
- 2517. Definitions in Mechanical Engineering.
- 2634. Pt. 1.
- 3027. Dimensions for Worm Gearing.
- 3643. 1963. ISO metric threads.

Handbook for Workshop Practice, No. 2.

British Standards Sales Branch, 2 Park Street, London, W.1.

Steel Tubes. Press. to 350 lb/in^2

Bore (in.)	Diam. (in.)	Wt/ft lb
$\frac{3}{4}$	$1\frac{1}{16}$	1·413
1	$1\frac{11}{32}$	2·023
$1\frac{1}{2}$	$1\frac{29}{32}$	3·837
2	$2\frac{3}{8}$	4·898
3	$3\frac{1}{2}$	8·678
4	$4\frac{1}{2}$	11·348
5	$5\frac{1}{2}$	14·019
6	$6\frac{1}{2}$	16·689
9	$9\frac{1}{2}$	30·667

Pipe Flanges (sizes in inches). Steam pressures up to 100 lb/in^2

Bore	Ext. Diam.	Flange Diam.	P.C.D.	Bolts No. Diam.		Flange Thickness C.I.	Steel
$\frac{1}{2}$	$\frac{27}{32}$	$3\frac{3}{4}$	$2\frac{5}{8}$	4	$\frac{1}{2}$	$\frac{1}{2}$	$\frac{3}{8}$
$\frac{3}{4}$	$1\frac{1}{16}$	4	$2\frac{7}{8}$	4	$\frac{1}{2}$	$\frac{1}{2}$	$\frac{3}{8}$
1	$1\frac{11}{32}$	$4\frac{1}{2}$	$3\frac{1}{4}$	4	$\frac{1}{2}$	$\frac{1}{2}$	$\frac{3}{8}$
$1\frac{1}{2}$	$1\frac{29}{32}$	$5\frac{1}{4}$	$3\frac{7}{8}$	4	$\frac{1}{2}$	$\frac{5}{8}$	$\frac{1}{2}$
2	$2\frac{3}{8}$	6	$4\frac{1}{2}$	4	$\frac{5}{8}$	$\frac{3}{4}$	$\frac{9}{16}$
3	$3\frac{1}{2}$	$7\frac{1}{4}$	$5\frac{3}{4}$	4	$\frac{5}{8}$	$\frac{3}{4}$	$\frac{9}{16}$

For further details refer to B.S. 10:1963.

Pipe Threads (sizes in inches)

Bore	Ext. Diam.	Threads		T.P.I.
		Top Diam.	Bottom Diam.	
$\frac{1}{2}$	$\frac{27}{32}$	0·825	0·734	14
$\frac{3}{4}$	$1\frac{1}{16}$	1·041	0·950	14
1	$1\frac{11}{32}$	1·309	1·193	11
$1\frac{1}{2}$	$1\frac{29}{32}$	1·882	1·766	11
2	$2\frac{3}{8}$	2·347	2·231	11
3	$3\frac{1}{2}$	3·460	3·344	11

For further details refer to B.S. 21:1962.

Pipe Flanges (sizes in inches). Pressures up to 450 lb/in^2

Nominal Bore Pipe	Ext. Diam. Pipe	Flange Diam.	Bolt P.C.D.	Bolts No. Diam.		Flange Thickness
$\frac{1}{2}$	$\frac{27}{32}$	$4\frac{1}{2}$	$3\frac{1}{4}$	4	$\frac{5}{8}$	$\frac{3}{4}$
$\frac{3}{4}$	$1\frac{1}{16}$	$4\frac{1}{2}$	$3\frac{1}{4}$	4	$\frac{5}{8}$	$\frac{3}{4}$
1	$1\frac{11}{32}$	5	$3\frac{3}{4}$	4	$\frac{5}{8}$	$\frac{7}{8}$
$1\frac{1}{2}$	$1\frac{29}{32}$	6	$4\frac{1}{2}$	4	$\frac{3}{4}$	1
2	$2\frac{3}{8}$	$6\frac{1}{2}$	5	8	$\frac{5}{8}$	1
3	$3\frac{1}{2}$	8	$6\frac{1}{2}$	8	$\frac{5}{8}$	$1\frac{1}{4}$

For further details refer to B.S. 10:1962.

HYDRAULIC COUPLINGS

Hydraulic Couplings (sizes in inches). Pressure 1,500 lb/in^2

Tube		Flange (oval and rectangular)					
Bore	Diam.	L	W	Th.	P.C.D.	Bolts	No.
$\frac{1}{4}$	$\frac{17}{32}$	$3\frac{1}{4}$	$1\frac{1}{2}$	$\frac{5}{8}$	2	$\frac{1}{2}$	2
$\frac{3}{8}$	$\frac{11}{16}$	$3\frac{1}{4}$	$1\frac{3}{4}$	$\frac{3}{4}$	$2\frac{1}{8}$	$\frac{1}{2}$	2
$\frac{1}{2}$	$\frac{27}{32}$	$4\frac{1}{4}$	2	$\frac{7}{8}$	$2\frac{3}{4}$	$\frac{5}{8}$	2
$\frac{3}{4}$	$1\frac{1}{16}$	$4\frac{1}{4}$	$2\frac{1}{4}$	1	$2\frac{7}{8}$	$\frac{5}{8}$	2
1	$1\frac{11}{32}$	$5\frac{1}{4}$	$2\frac{3}{4}$	$1\frac{1}{8}$	$3\frac{1}{2}$	$\frac{3}{4}$	2
$1\frac{1}{2}$	$1\frac{29}{32}$	6	$3\frac{1}{2}$	$1\frac{3}{8}$	$4\frac{1}{4}$	$\frac{7}{8}$	2
2	$2\frac{3}{8}$	7	4	$1\frac{3}{4}$	5	1	2

See No. 331.

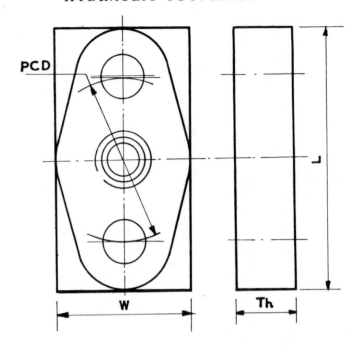

Hydraulic Couplings (sizes in inches). Pressure 1,500 lb/in^2

Tube		Flange (Square)				
Bore	Diam.	L	Th.	P.C.D.	Bolts	No.
$\frac{1}{4}$	$\frac{17}{32}$	$2\frac{3}{4}$	$\frac{3}{4}$	2	$\frac{1}{2}$	4
$\frac{3}{8}$	$\frac{11}{16}$	$2\frac{3}{4}$	$\frac{3}{4}$	$2\frac{1}{8}$	$\frac{1}{2}$	4
$\frac{1}{2}$	$\frac{27}{32}$	$2\frac{3}{4}$	$\frac{7}{8}$	$2\frac{1}{4}$	$\frac{1}{2}$	4
$\frac{3}{4}$	$1\frac{1}{16}$	3	$\frac{7}{8}$	$2\frac{1}{2}$	$\frac{1}{2}$	4
1	$1\frac{11}{32}$	$3\frac{3}{4}$	1	$3\frac{1}{4}$	$\frac{5}{8}$	4
$1\frac{1}{2}$	$1\frac{29}{32}$	$4\frac{1}{2}$	$1\frac{1}{8}$	4	$\frac{3}{4}$	4
2	$2\frac{3}{8}$	5	$1\frac{1}{8}$	$4\frac{1}{2}$	$\frac{3}{4}$	4
3	$3\frac{1}{2}$	$6\frac{1}{2}$	$1\frac{3}{4}$	$6\frac{1}{4}$	1	4

See No. 332.

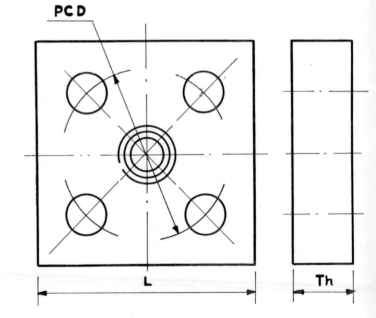

262

Standard Pitches

Spur Gears

Diametral Pitch (in.)	Circular Pitch (in.)	Module (mm)
D.P. (P)	C.P. (p)	M. (m)
16	0·1963	1·587
12	0·262	2·118
10	0·314	2·540
8	0·393	3·177
6·283	0·5	4·042
6	0·524	4·236
5	0·628	5·077
4	0·785	6·350
3·142	1·000	8·085
3	1·047	8·465
2·5	1·257	10·163
2	1·571	12·701
1·5	2·094	16·930
1	3·142	25·400

$$\text{D.P.} = \frac{T}{\text{P.C.D.}} \qquad \text{P.C.D.} = \frac{T}{\text{D.P.}}$$

$$\text{C.P.} = \frac{\pi \times \text{P.C.D.}}{T} \qquad M = \frac{\text{P.C.D.}}{T}$$

Pressure Angle = 20°

Addendum = 0·3183 × C.P.

Dedendum = 0·3683 × C.P.

$$\text{Clearance} = \frac{\text{C.P.}}{20}$$

B.S. 436 : 1940.

Spur Gears

1″ Circular Pitch

No. of Teeth	P.C.D. (in.)
10	3·183
11	3·501
12	3·820
13	4·138
14	4·456
15	4·775
16	5·093
18	5·730
20	6·366
22	7·003
24	7·639
26	8·276
28	8·913
30	9·549
32	10·186
36	11·459

(See Nos. 113 to 123.)

Sunk Keys (in inches)

Shaft ϕ	W
$\frac{3}{8}$	$\frac{1}{8}$
$\frac{1}{2}$	$\frac{5}{32}$
$\frac{3}{4}$	$\frac{3}{16}$
1	$\frac{1}{4}$
$1\frac{1}{4}$	$\frac{5}{16}$
$1\frac{1}{2}$	$\frac{3}{8}$
$1\frac{3}{4}$	$\frac{7}{16}$
2	$\frac{1}{2}$
3	$\frac{3}{4}$

B.S. 46 (see 294.)

Woodruff Keys

No.	W	D	E	F
204	$\frac{1}{16}$	$\frac{1}{2}$	0·203	0·167
404	$\frac{1}{8}$	$\frac{1}{2}$	0·203	0·136
606	$\frac{3}{16}$	$\frac{3}{4}$	0·313	0·214
808	$\frac{1}{4}$	1	0·438	0·308
1010	$\frac{5}{16}$	$1\frac{1}{4}$	0·547	0·386
1212	$\frac{3}{8}$	$1\frac{1}{2}$	0·641	0·448

B.S. 46. (see No. 294.)

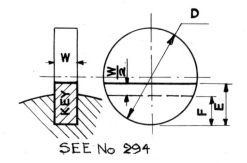

SEE No 294

263

Limits and Fits

Hole	Shafts Average Running	Shafts Average Location	Shafts Keying	Shafts Heavy Press	Hole	Shafts Precision Running	Shafts Precision Location	Shafts Keying	Shafts Press (Ferrous)	Shafts Press (Non-Ferrous)	Nominal Diam. (in.)
H8	f8	h7	j7	s7	H7	f7	g6	k6	p6	s6	over-to
+0·9 +0	−0·5 −1·4	−0 −0·6	+0·4 −0·2	+1·6 +1·0	+0·6 +0	−0·5 −1·1	−0·2 −0·6	+0·5 +0·1	+1·0 +0·6	+1·4 +1·0	0·24–0·40
+1·0 +0	−0·6 −1·6	−0 −0·7	+0·5 −0·2	+1·9 +1·2	+0·7 +0	−0·6 −1·3	−0·25 −0·65	+0·5 +0·1	+1·1 +0·7	+1·6 +1·2	0·40–0·71
+1·2 +0	−0·8 −2·0	−0 −0·8	+0·5 −0·3	+2·2 +1·4	+0·8 +0	−0·8 −1·6	−0·3 −0·8	+0·6 +0·1	+1·3 +0·8	+1·9 +1·4	0·71–1·19
+1·6 +0	−1·0 −2·6	−0 −1·0	+0·6 −0·4	+2·8 +1·8	+1·0 +0	−1·0 −2·0	−0·4 −1·0	+0·7 +0·1	+1·6 +1·0	+2·4 +1·8	1·19–1·97

Tolerances in Units of 0·001 in.

For further details refer to B.S. 1916: Parts 1 and 2: 1953.

Newall Tolerances

Nominal Diam. (in.)	Hole	Shafts Running X	Shafts Running Y	Shafts Running Z	Push	Drive	Force
to ½	A +0·00025 −0·00025	−0·001 −0·002	−0·00075 −0·00125	−0·00050 −0·00075	−0·00025 −0·00075	+0·00050 +0·00025	+0·0010 +0·0005
	B +0·0005 −0·0005						
9/16 to 1	A +0·00050 −0·00025	−0·00125 −0·00275	−0·0010 −0·0020	−0·00075 −0·00125	−0·00025 −0·00075	+0·0010 +0·00075	+0·0020 +0·0015
	B +0·00075 −0·00050						
1 1/16 to 2	A +0·00075 −0·00050	−0·00175 −0·00350	−0·00125 −0·00250	−0·00075 −0·00150	−0·00025 −0·00075	+0·0015 +0·0010	+0·004 +0·003
	B +0·0010 −0·0005						

Angular Measure

Degrees	Radians	Radians	Degrees
1	0·0175	0·01	0·57
2	0·0349	0·02	1·15
3	0·0524	0·03	1·72
4	0·0698	0·04	2·29
5	0·0873	0·05	2·86
10	0·1745	0·10	5·73
20	0·3491	0·20	11·46
30	0·5236	0·50	28·65
60	1·047	0·75	42·97
90	1·571	1·00	57·30
120	2·094	1·50	85·94
150	2·618	$\pi/2$	90·00
180	3·142	π	180·00
360	2π	2π	360·00

1 Radian = $\dfrac{180}{\pi}$ degrees = 57·29578°.

1 Degree = 0·0174533 radians.

Angular Velocity

One radian per sec = 9·5493 r.p.m. = 57·296° per sec.
One degree per sec = 0·16667 r.p.m. = 0·017453 radians per sec.
One r.p.m.   = 6° per sec   = 0·10472 radians per sec.

Material	Composition %	Tension ton/in^2
Cast Iron	C 0·6	10 to 15
Cast Iron	C 3, Mn 0·8, Si 2·2	18 to 22
Cast Iron, Malleable	C 3, Mn 1·0, P 0·1, Ni 0·3, Mo 1·0	35 to 40
Steel, Mild	C 0·1, Mn 1·0	23 to 27
Steel, Cast	C 0·4, Mn 0·8,	40 to 45
Steel, Nickel	C 0·3, Mn 0·5, Ni 3, Cr 0·8, Mo 0·5	50 to 65
Steel, Stainless	C 0·25, Mn 1·0, Ni 2, Cr 16·0, Si 0·5	55
Aluminium Alloy	Al 84, Zn 13, Cu 3	11
Aluminium Bronze	Cu 90, Al 10	50
Duralumin	Al 95·5, Cu 4, Mg 0·5	27
Magnesium Alloy	Mg 91·3, Al 8, Zn 0·5, Mn 0·2	20
Brass, Std.	Cu 70, Zn 30	25
Brass, Turning	Cu 58, Zn 40, Pb 2	15
Brass, Pressing	Cu 60, Zn 40	19
Gunmetal	Cu 88, Sn 10, Zn 2	16
Copper Nickel	Cu 80, Ni 20	—
Monel Metal	Cu 30, Ni 67, Fe 1, S 1	—
Babbit No. 1	Sn 83·3, Cu 8·33, Sb 8·33	—
Lead Bronze	Cu 74, Pb 25, Sn 1·2	—
Phosphor Bronze	Cu 89·5, Sn 10, P 0·5	20 to 30

Key: C=carbon; Al=aluminium; Cr=chromium; Cu=copper; Fe=iron; Mg=magnesium; Mn=manganese; Mo=molybdenum; Ni=nickel; P=phosphorus; Pb=lead; S=sulphur; Sb=antimony; Si=silicon; Sn=tin; Zn=zinc.

APPENDIX II

1. Two thin walled tubes 3″ diam. and 2″ diam. respectively are arranged as shown in Fig. 1 with their axes parallel. When the larger tube rolls on the horizontal plane, it may be assumed that the smaller tube rolls inside it so that the axes of both tubes remain in the same vertical plane. If the larger tube rolls clockwise through one-third of a revolution, draw the locus of a point A on the inner tube:

(a) with respect to the horizontal plane,
(b) with respect to the outer tube.
. (U.L.)

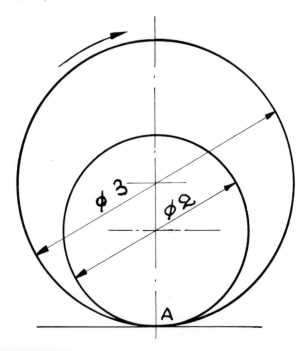

fig. 1

2. The asymptotes of a hyperbola are shown in Fig. 2. Construct:

(a) one branch of a hyperbola to pass through point P,
(b) one branch of the conjugate hyperbola.

Draw on the diagram the two auxiliary circles and mark clearly all foci and directrices of the hyperbola and its conjugate.
(A.E.B.)

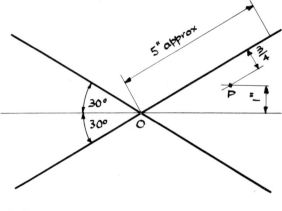

fig. 2

Cams

3. Fig. 3 shows a cam and $1\frac{1}{4}''$ diam. roller follower on an oscillating arm, which is shown horizontal and in the lowest position. The arm moves the slider block along a vertical path. The slider is required to remain stationary during 30° of rotation of the cam, then to move 3″ with SMH (rest to rest) along the slide path during 90° rotation of the cam. The slider is then to be at rest during 90° rotation of the cam and to return with SMH to its lowest position, during the remainder of rotation.

Construct, half full size, the profile of the uniformly rotating cam. Use a clockwise direction of rotation.
(N.U.J.M.B.)

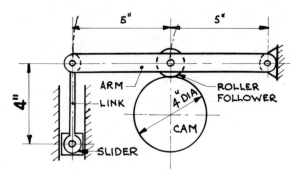

fig. 3

4. A cam has a minimum radius of $1\frac{1}{4}''$ and a maximum radius of $2\frac{1}{2}''$. Motion is transmitted to a roller follower of diam. 1″ which is constrained to move in a straight line offset $\frac{3}{4}''$ from the centre-line of the cam as shown in Fig. 4.

The follower rises with simple harmonic motion for 180° of cam rotation and falls with simple harmonic motion for the next 90° of cam rotation; there is a dwell at the minimum radius for the final 90° of cam rotation.

Construct, full size, the profile of the cam.
(A.E.B.)

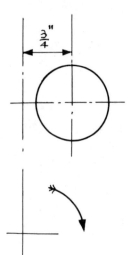

fig. 4

Helices

5. A right-hand helical spring is made from wire of diam. $\frac{1}{2}''$. There are one-and-three-quarter turns of mean diam. $2''$ and lead (pitch) $2''$. The ends of the spring are ground square to the axis of the spring.

Draw the plan and elevation of the spring standing with its axis vertical. One end of the helical centreline of the wire is to be at the bottom left-hand side of the elevation. The shape of the concealed end of the wire must be shown on the plan as a dotted line.
(A.E.B.)

Frameworks

6. As shown in Fig. 5, a vertical load of 2,000 lb wt. is winched by a horizontal cable which passes over a pulley mounted at the end of the pin-jointed cantilever frame. Construct the force polygon and neglecting the weight of the structure find the reactions and the magnitude and nature of the force in each member. Tabulate your answers.

Use a scale of $1''$ represents 500 lb wt.
(J.M.B.)

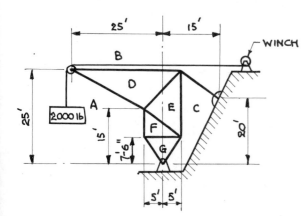

fig. 5

7. A pin jointed structure, loaded as shown in Fig. 6 is hinged at B to a fixed pivot. The joint at A is supported on rollers. Find the magnitude and direction of the reactions A and B.

Draw a force diagram to a scale $1'' = 500$ lb. From your diagram evaluate the forces in the tension members.
(U.L.)

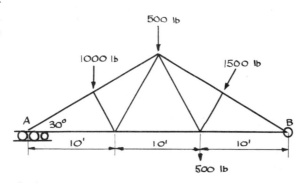

fig. 6

8. A cantilever is acted upon by the forces shown in Fig. 7. Draw:

(a) the space diagram full size,
(b) the force polygon using a scale of $1''$ to 2 lbf,
(c) the bending moment diagram using a polar distance of $2''$.

Show clearly on the space diagram the magnitude and direction of the reaction force and couple at the wall.
(A.E.B.)

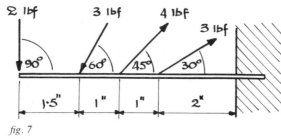

fig. 7

Centroids

9. For the symmetrical section shown in Fig. 8 determine graphically:

(a) the position of the centroid,
(b) the second moment of the area about the axis XX.
(A.E.B.)

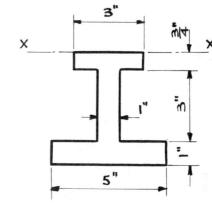

fig. 8

Mechanisms

10. Fig. 9 shows a form of straight line mechanism. When C is at the mid-position as shown, AO_1 is vertical and ABC lie in a horizontal line.

Using a scale of twice full size and by plotting a sufficient number of points, show that the locus of the point C is a straight line when the piston P moves up and down $0.25''$.

State in inches the travel of P when C moves $\pm 1''$ from its mid-position.
(A.E.B.)

$AO_1 = 2''$
$PD = 1''$
$BO_2 = 1''$
$BD = \frac{1}{2}''$
$BC = 2''$

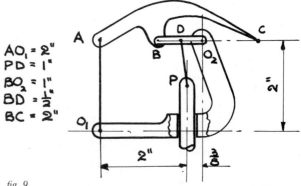

fig. 9

267

11. Five gears each of 4″ pitch circle diameter, mesh together and rotate about fixed axes in the manner shown in Fig. 10. Opposite gears are coupled by the connecting rods. The central gear rotates through one revolution from the initial position, thus driving the meshing gears.

Draw, full size, and plot the locus of the intersection of the connecting rods:

(*a*) relative to space,
(*b*) as traced on the central gear.
(J.M.B.)

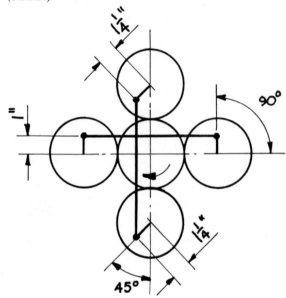

fig. 10

12. A single-slider-crank-chain is shown in Fig. 11. Draw, full size, the two following inversions of the mechanism,

(*a*) when A B is the fixed link and B C rotates,
(*b*) when B C is the fixed link and A B rotates.

In each case draw,
(i) the two positions of the mechanism at angles of crank rotation of 30° and 210°,
(ii) the two extreme positions of the mechanism.
Identify each point A, B, C and D, and show the angles 30° and 210° from the chosen axis of reference.
(A.E.B.)

268

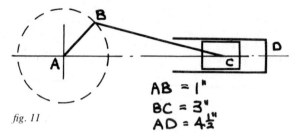

fig. 11

AB = 1″
BC = 3″
AD = 4½″

Inclined Planes

13. A ring of circular cross section is shown in Fig. 12. Draw the given view and add the following views:

(*a*) a sectional elevation as seen from the left-hand side, and
(*b*) a sectional plan.

The plane of the section is to be taken on X X.
(U.L.)

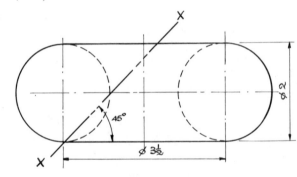

fig. 12

14. A right triangular pyramid with sides of 3″ in true length at the base, is shown in the plan view of Fig. 13. The base lies in the plane V T H with one edge of the base in the ground plane. The pyramid is cut by a second plane $V^1T^1H^1$.

Indicating any hidden lines, draw full-sized views in the V.P. and H.P. together with an end elevation, of the part of the pyramid between the two planes V T H and $V^1T^1H^1$.

Develop all the surfaces of this part including those lying in the planes V T H and $V^1T^1H^1$.
(J.M.B.)

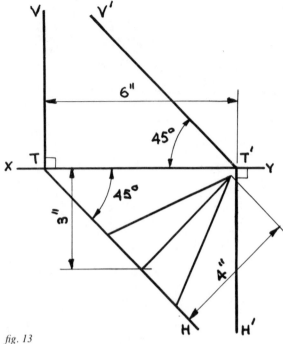

fig. 13

Oblique Planes

15. The cylindrical pipe shown in Fig. 14 is held eccentrically in a lathe chuck so that it rotates about axis O O. Metal is removed by taper turning along the line T T.

Draw, full size, three orthographic views including one in the given plane, of the solid which results from the machining operation.
(J.M.B.)

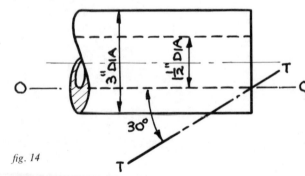

fig. 14

16. Fig. 15 shows the plan and elevation of an oblique cone. The base A B has a diameter of 3″, the length B C is 5″ and A C is 3½″. All sections parallel to the base are circular.

The cone is intersected by an oblique plane shown by the traces VT and HT.

Draw, full size, the elevation and plan of the intersected cone.

(A.E.B.)

Oblique Planes

17. In Fig. 16 the line E F is parallel to the plane surface A B C D. Complete the elevation of the plane surface and find its true shape.

(U.L.)

18. The projections of the two lines A B and C D are given in Fig. 17. Draw the projections of a plane square surface with its centre at B and one edge contained by the line C D. (C D is not an edge of the square.) Determine the angle which the line A B makes with the surface.

(U.L.)

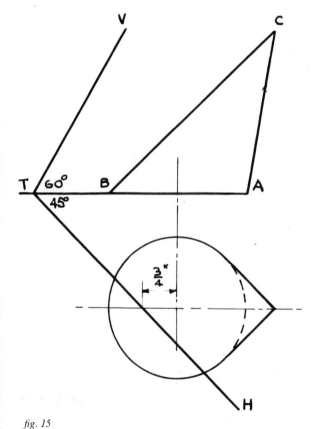

fig. 15

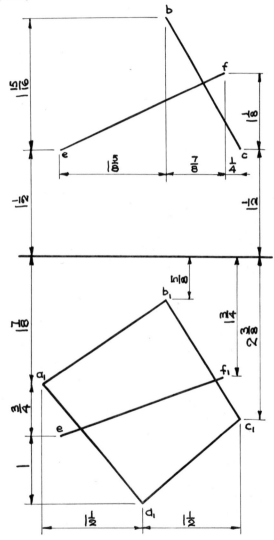

fig. 16

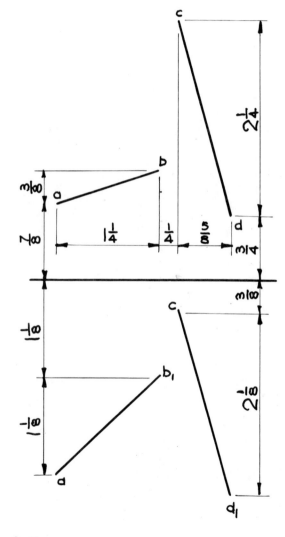

fig. 17

269

Solid Geometry
Interpenetrations

19. Fig. 18 shows incomplete views of a hemisphere and intersecting cylinder. Complete these views showing the curves of intersection. Indicate all the hidden lines. (U.L.)

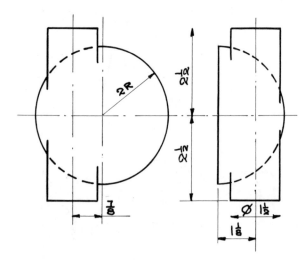

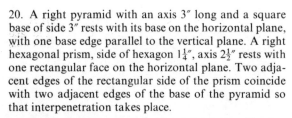

fig. 18

20. A right pyramid with an axis 3″ long and a square base of side 3″ rests with its base on the horizontal plane, with one base edge parallel to the vertical plane. A right hexagonal prism, side of hexagon $1\frac{1}{4}$″, axis $2\frac{1}{2}$″ rests with one rectangular face on the horizontal plane. Two adjacent edges of the rectangular side of the prism coincide with two adjacent edges of the base of the pyramid so that interpenetration takes place.

(a) Draw the plan and elevation of the figures showing the lines of intersection, including the hidden lines.
(b) Draw the development of that part of the prism which protrudes from the pyramid.
(A.E.B.)

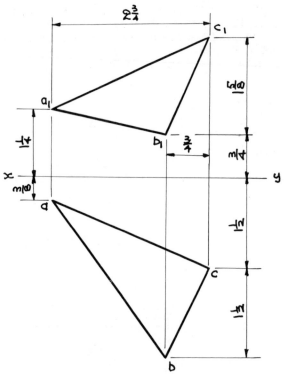

fig. 19

Lines of True Length

21. The plan and elevation of a triangle ABC are shown in Fig. 19. Determine and state:

(a) The true lengths of the sides of the triangle.
(b) The angles between the ground line xy and the horizontal and vertical traces of the oblique plane in which the triangle lies.
(c) The true angle between the plane of the triangle and the horizontal plane.
(A.E.B.)

22. Develop the surface of the open-ended sheet metal duct shown in Fig. 20 which takes the form of a triangular prism, and also determine the angles between the sides.
(U.L.)

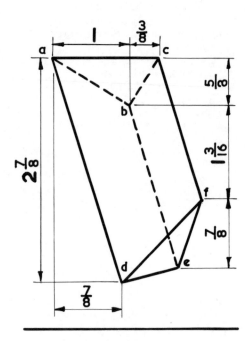

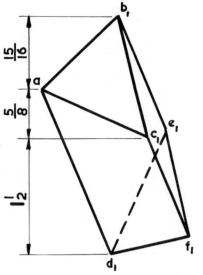

fig. 20

270

Isometric

23. Details of a bracket are given in Fig. 21. Draw full size an isometric view arranged so that the spot-faced boss is near the viewer. Lines representing hidden parts need not be shown.

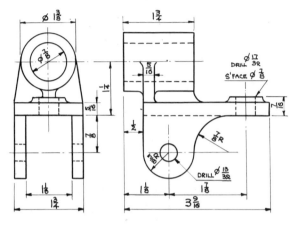

fig. 21

24. Details of a handrail bracket are given in Fig. 22. Using a scale of full size, draw an isometric view of the bracket, with the curved portion marked O in the lower right-hand corner of your drawing. Hidden lines are not required.
(U.L.)

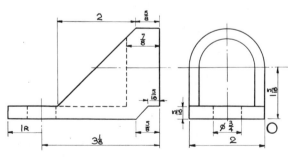

fig. 22

25. Draw, full size, an isometric scale.

Use this scale to produce an isometric view of the object shown in Fig. 23 viewed in the direction of the arrows. Hidden lines must be shown representing the edges of the pyramid.

Do not show the hidden line representing the curve of

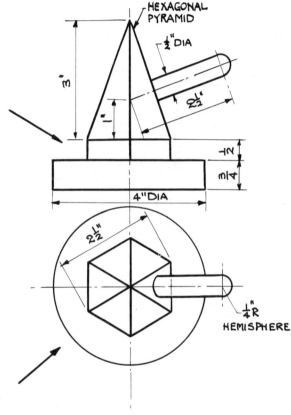

fig. 23

intersection of the $\frac{1}{2}''$ diam. cylinder and the side of the pyramid.
(A.E.B.)

Sketching Questions

26. Two pictorial views of an aluminium alloy casting are shown, in the figure. The dovetail slide is machined.

Make a three-plane orthographic sketch, in good proportion, using any aids you wish. Show fillets, external radii, and machining marks where necessary. Select views on which you can most clearly show the dimensions and instructions required for pattern and machine shop use. Make improvements in dimensioning where you consider these necessary.
(J.M.B.)

27. Make neat sketches of one of the following:

(*a*) an air cleaner for a motor car engine, or
(*b*) an exhaust silencer for a vehicle.

The views must include a sectional view to illustrate the method of assembly and operation (freehand, on squared paper). Orthogonal views, not pictorial.
(U.L.)

28. Make neat sketches, including a sectional view, of *one* of the following:

(*a*) a float operated valve,
(*b*) a simple safety valve, or
(*c*) a non-return valve.
(U.L.)

29. It is common to use two way switching for a light at the bottom and another at the top of a flight of stairs.

Draw, freehand, a typical wiring diagram from the fuse to a junction box and then on to the two pairs of switches needed to operate the two lights independently.
(A.E.B.)

30. Throughout the country there are many steel pylons which support high voltage electric cables.

Make freehand sketches to show:

(*a*) typical cross sections of the steel used;
(*b*) an isometric view of part of the structure to show how the sections are fastened together to produce the necessary rigidity.
(A.E.B.)

31. Draw, freehand, isometric or orthographic views to show the construction and assembly of the leaf spring of a car or a lorry. Give separate details of the U-bolts and shackle pins.
(A.E.B.)

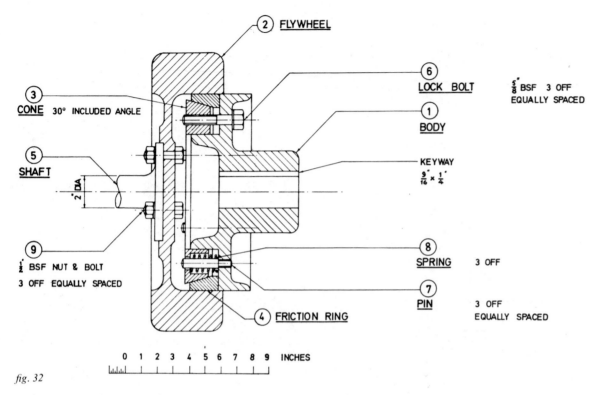

② FLYWHEEL

③ CONE 30° INCLUDED ANGLE

⑤ SHAFT

2″ DIA

⑨
½″ BSF NUT & BOLT
3 OFF EQUALLY SPACED

⑥ LOCK BOLT

⅜″ BSF 3 OFF
EQUALLY SPACED

① BODY

KEYWAY
$\frac{9}{16}$″ × ¼″

⑧ SPRING 3 OFF

⑦ PIN 3 OFF
EQUALLY SPACED

④ FRICTION RING

0 1 2 3 4 5 6 7 8 9 INCHES

fig. 32

33. The drawing Fig. 33 shows the detailed components of a right-angled pipe valve. All parts except the valve are detailed completely.

Draw in Third Angle Projection the following views of the assembled valve:

(a) a sectional elevation on line X X,
(b) the end elevation as viewed from arrow A.

Design and draw the valve end B to suit the main valve body. The valve should be shown in the fully-closed position. Hidden lines are not required in either view.

Suitable nuts and bolts should be shown where necessary. Show a suitable means of attaching the handwheel to the valve stem. List on your drawing the ten dimensions which you think would be of most interest to an engineer wishing to fit such a valve into a pipe line.

3 hours allowed.

(J.M.B.)

32. The diagram Fig. 32 shows, to the scale indicated, the axial section of a friction coupling for connecting two co-axial shafts of 2″ diam.

The body, detail 1, can be attached to the right-hand shaft by means of a B.S. rectangular key and the flywheel, detail 2, is bolted to a flange on the shaft, detail 5. Details 1 and 2 can be locked together by details 3 and 4. Detail 3, which is conical, fits inside detail 4.

Detail 4 is split radially by a $\frac{1}{16}$″ wide saw-cut at a suitable point on its circumference and is an easy fit inside the flywheel which is machined to an inside diameter of 12″.

When assembled the coupling is locked by tightening the three bolts, detail 6; loosening the bolts a turn will release the coupling for easy separation of the shafts. The three pins and springs, details 7 and 8, act as extractors.

(a) Draw (in orthographic projection), three-eights full size, the following views of details 1, 2, 3 and 4 showing hidden detail:
 (i) an elevation corresponding to the given view, the top half to be in section and the bottom half an outside view;
 (ii) an outside end elevation.
(b) Draw, three-eighths full size, two outside views of details 5, 6, and 7.
(c) Draw, three-eighths full size, two outside views of the spring, detail 8, using B.S. representation. The free length of the spring is $2\frac{3}{4}$″ and the wire diameter is $\frac{1}{8}$″.
(d) Fully dimension each detail and add the name of each in letters $\frac{3}{16}$″ high.
(e) Add in letters $\frac{3}{8}$″ high the title FRICTION COUPLING.

272

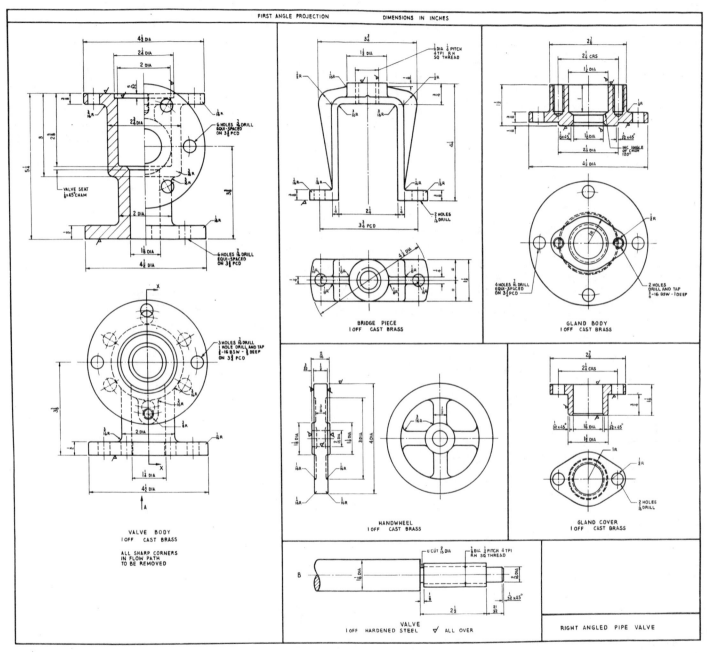

VALVE SEAT
⅛ x 45° CHAM

VALVE BODY
1 OFF   CAST BRASS

ALL SHARP CORNERS
IN FLOW PATH
TO BE REMOVED

BRIDGE PIECE
1 OFF   CAST BRASS

GLAND BODY
1 OFF   CAST BRASS

HANDWHEEL
1 OFF   CAST BRASS

GLAND COVER
1 OFF   CAST BRASS

VALVE
1 OFF HARDENED STEEL   ∀ ALL OVER

RIGHT ANGLED PIPE VALVE

fig. 33

273

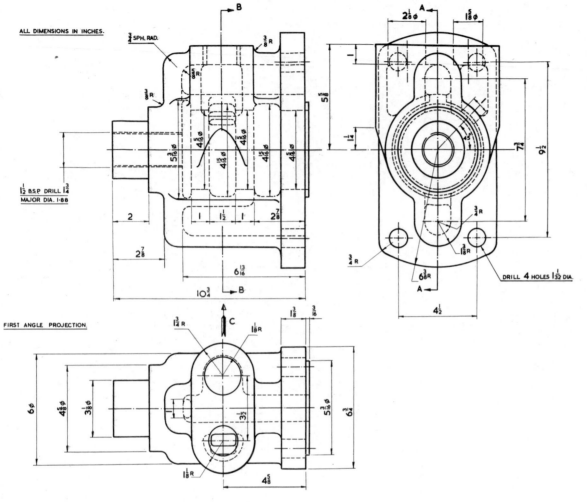

ALL DIMENSIONS IN INCHES.

FIRST ANGLE PROJECTION.

34. The diagram Fig. 34 shows views of a Valve Body Casting.

Draw, to a scale of three-quarters full size:

(a) A sectional elevation on AA in the direction of the arrows.

(b) A half section on BB in the direction of the arrows omitting that portion to the right of the main centre line.

(c) A complete view in the direction C projected above. Dotted lines representing hidden parts are not required. Small radii may be taken as $\frac{1}{8}''$ and any missing dimensions should be estimated.

3 hours allowed.

(U.L.)

fig. 34

274

Index

The numbers in the Index refer to sections in the text.